BK 541.39   D274U
UNDERSTANDING CHEMICAL REACTIONS
                                    /DAY, M C
    C1986   15.00   FV

3000 717204 30010
St. Louis Community College

D0946834

WITHDRAWN

541.39 D274u   F V
DAY
  UNDERSTANDING CHEMICAL
  REACTIONS
                                    15.00

St. Louis Community College

Library

5801 Wilson Avenue
St. Louis, Missouri 63110

UNDERSTANDING CHEMICAL REACTIONS

M. C. Day

Louisiana State University

Barry Corona

Spring Hill College

Allyn and Bacon, Inc.

Boston London Sydney Toronto

To Gloria

Copyright © 1986 by Allyn and Bacon, Inc., 7 Wells Avenue, Newton, Massa-
chusetts 02159. All rights reserved. No part of the material protected by this
copyright notice may be reproduced or utilized in any form or by any means,
electronic or mechanical, including photocopying, recording, or by any information storage or retrieval system, without written permission from the copyright owner.

Cover Coordinator: Linda Dickinson

Library of Congress Cataloging-in-Publication Data

Day, M. C.
   Understanding chemical reactions.

   1. Chemical reactions. I. Corona, Barry.
II. Title.
QD501.D34 1986   541.3'9   85-28780
ISBN 0-205-08607-1

Printed in the United States of America

10 9 8 7 6 5 4 3 2 1   91 90 89 88 87 86

Contents

Preface

For close to thirty years after World War II, the emphasis in chemistry at the freshman level was oriented toward theoretical principles at the expense of descriptive chemistry. This change in emphasis was, undoubtedly, justified with the logic being (a) if one knows the principles, he or she can figure out the answers, and (b) one can always use a reference book. In retrospect, we can see we went too far, and now the mood seems to be shifting back as attested to by the large amount of descriptive chemistry found in many of the more recent freshman texts. Perhaps this shift will also go too far, but certainly the total lack of familiarity with descriptive chemistry observed among entering graduate students, as well as with many of us from earlier generations, speaks rather eloquently to the previous lack of balance.

It is much more likely that we have now had our experience with extremes and will teach the material with an appreciation of the importance of both descriptive and theoretical chemistry. It is certainly a truth that, in the future, a student will be expected to know some descriptive chemistry, with chemical reactions being an integral part of that chemistry. It is to this latter problem, understanding chemical reactions, that this book is directed.

It is interesting that our experience with teaching descriptive chemistry, and chemical reactions in particular, has often been discouraging. No matter how well the material seems to be presented in the text, the topic is a difficult one for the student. Based on our contacts with large freshman classes, conversations with each other and our colleagues and, most importantly, with our students, we have tried to determine the sources of the problem. It is our conclusion that the usual presentation is fragmented, with little or no systemization. The consequence is that the student is overwhelmed by the trials of apparent random memory. It is out of this conviction that we have written this book.

In one sense, this is a programmed learning book, but in another sense, it is not. Rather than merely presenting organized questions and answers, we have tried to give a rational basis for the answers. Consequently, much of the material can stand on its own. Yet, in general, we assume that the student has been exposed to the usual topics in the classroom. In writing this book, we have had the professors as well as the students in mind. The chapter on nomenclature has been written in such a manner that the unprepared student may learn the names of common chemical substances without the necessity of a formal lecture. In contrast, a chapter such as the one on the periodic table may, at first, seem to be nothing more than a repeat of the standard lecture material. And perhaps it is. However, the material presented in this first chapter has been carefully selected. It is that part of the usual presentation of the periodic table that is of particular importance in predicting the products of chemical reactions. We strongly urge that this chapter not be skipped by the student.

The general structure of the book is not arranged to supplement any particular freshman text. Rather, chapters can be selected in the order that best fits a given coursework plan. Additionally, in some cases, the level of presentation may go beyond what is normally taught in an introductory course. A discussion of *meta-*, *meso-*, and *ortho-* forms of acids and polyatomic anions, for example, may fall into this category. Textbooks often mention these forms without explaining their origin. We feel that an explanation is in order. Others may not agree, and certainly they should feel free to omit it. It is not our intention that every student necessarily read every page and work every problem. The student, along with the professor's guidance, should read the sections and work the problems that meet his or her needs.

Whereas this book was written primarily with the introductory chemistry course in mind, the authors feel that it will be of help both as a review for students taking an advanced inorganic chemistry course and as a review of basic chemistry for new graduate students. We also encourage high school teachers to use it either as a review or as a learning tool, whichever their need may be.

Throughout, we have attempted to show only reactions that do, in fact, actually occur, except where we have specifically stated the contrary. If any errors have been made, we would appreciate being so informed.

There have been many who have contributed to the completion of this manuscript. We want to particularly express our appreciation to Professors Dewey Carpenter, Paul Koenig, and Buddhadev Sen for their helpful comments on our first draft. We further wish to acknowledge Diane McPherson and Linda Corona for typing parts of the manuscript and especially wish to thank Angela Burke, who typed several chapters under both a deadline and trying circumstances.

USE OF THIS BOOK

It is rather commonplace for a student to be convinced that he or she understands certain material because of a particularly lucid lecture or from reading, usually repetitiously, the material in the text. The subsequent performance on a test, more likely than not, proves the contrary to be the case. The truth is, chemistry is best studied with a pencil and paper. In studying chemistry, it is also important that the student knows at that time whether a particular answer is correct. If it is not, he or she had better find out why. This book incorporates such a pencil-and-paper approach to chemistry.

You will find many exercises throughout the book with the answers at the back. The answer sheets are perforated so that they can be removed. To make the most effective use of this book, you should write down the answers to the questions and check your responses with the answer sheet. If you miss a question, it is essential that you restudy the appropriate material until you understand why the correct response is, in fact, correct. Sometimes, it may turn out that even after considerable study, you cannot determine the reason for a particular response. In such instances you should go on to the next question and discuss the problem with your professor at the earliest opportunity.

The conventional manner for checking your responses is to place your answer sheet next to your responses and cover it with a piece of paper. As you answer a question, you correspondingly check your answer.

1

The Periodic Table

Chemical reactions are frequently very complicated, and it is not to be inferred from the title of this book that one can learn to predict the products of all chemical reactions. In fact, it would be difficult to correctly predict more than a small percentage of those known. Nevertheless, by learning a few simple principles and applying a little common sense, one can predict the results of a sizeable number of reactions. And even if one ends up with incorrect predictions, they should, at least, be reasonable.

For instance, let's consider the reaction between sodium and phosphoric acid. If we merely arrange atoms and numbers, it is possible to write out an unlimited list of products, such as

$$Na + H_3PO_4 \longrightarrow HNaP + O_2$$

or
$$\longrightarrow Na_7O_{13} + O_3P + H_7$$

or
$$\longrightarrow O_7P_{15} + H_{26}Na_3$$

and so on

But all of these proposed products are ridiculous. The question is *how does one know what products are or are not reasonable*? The purpose of this book is to give the beginning chemistry student some helpful guidelines as well as to serve as a review for those who have studied reaction chemistry too long ago. We are assuming that in your chemistry lecture you have already been introduced to the topics discussed here. But we have selected and arranged the material to emphasize those aspects of beginning chemistry that will be particularly helpful in predicting the products of chemical reactions.

ATOMIC STRUCTURE AND PERIODICITY

The starting point of any discussion of chemical reactions is the periodic table. Historically the periodic table was constructed on the basis of the similarities of both the chemical and the physical properties of the elements. Thus, it is a visual representation of experimentally observed properties and is, therefore, factual. Consequently, if the structure of the periodic table can be understood, it should be possible to know something about why the elements behave as they do. Today we attempt to understand these similarities in terms of the arrangements of the electrons in the atoms.

Using the common long form of the periodic table, it can be seen from Figure 1.1 that the table can be divided into four sections containing 2, 6, 10, and 14 elements in the various horizontal arrays (periods). Recalling

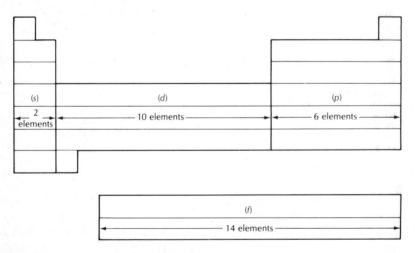

Figure 1.1 The structure of the long form of the periodic table in terms of s, p, d, and f blocks of electrons

from the electron configurations of the elements that the following relations exist,

$$s \text{ orbital} — 2 \text{ electrons}$$
$$p \text{ orbitals} — 6 \text{ electrons}$$
$$d \text{ orbitals} — 10 \text{ electrons}$$
$$f \text{ orbitals} — 14 \text{ electrons}$$

it should be apparent that the division of the periodic table into the four sections is related to the *s, p, d,* and *f* electrons in the atoms of the respective elements.

We will soon see the relationship between these sections in the periodic table and the electron arrangements in the individual atoms. But for now, let's see if you can correlate the elements with the four regions.

_____ **EXERCISE** _____

1.1 Using the periodic table, indicate the regions where the following elements occur. (Note that the atomic numbers are given to help you locate the different elements.)

a. $_{16}$S is in a region of the periodic table characterized by __*p*__ electrons.

b. $_{19}$K is in a region of the periodic table characterized by __*s*__ electrons.

c. $_{24}$Cr is in a region of the periodic table characterized by __*d*__ electrons.

d. $_{13}$Al is in a region of the periodic table characterized by __*p*__ electrons.

e. $_{15}$P is in a region of the periodic table characterized by __*p*__ electrons.

f. $_{63}$Eu is in a region of the periodic table characterized by __*f*__ electrons.

g. $_{47}$Ag is in a region of the periodic table characterized by __*d*__ electrons.

═══════════════════════════

Classification of the Elements. For convenience, it is possible to divide the elements into four classes based on rather broad similarities in properties. Although there are several ways that this may be done, one useful classification is shown in Figure 1.2.

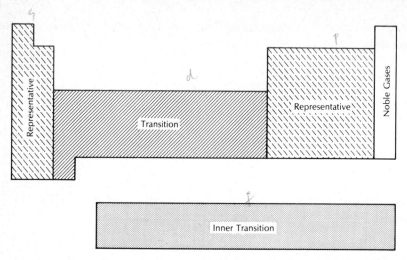

Figure 1.2   The four classes of elements in the periodic table

In this classification the *representative* elements are those filling in *s* or *p* electrons and, in most periodic tables, are listed as **A** group elements. The *transition* elements are filling in *d* electrons and are usually designated as **B** group and group VIII elements. The *noble gases,* which were referred to as *inert gases* for many years, are those elements, except for helium, that have a completed *p* level. Finally, the *inner transition* elements are those filling in *f* electrons. The inner transition elements can be further broken down to the *lanthanides* (elements 58–71) and the *actinides* (elements 89–103). The lanthanides are also referred to as the *rare earth* elements and the actinides are sometimes called the *second rare earth* series.

As mentioned above, there are other means of classifying the elements. For instance, one very common classification includes the noble gases with the representative elements. This is convenient in terms of electron configurations. However, historically, classification was based on general similarities in the behavior of the elements. On this basis, it is certainly justified to separate the noble gases from the representative elements.

—————— **EXERCISES** ——————

1.2   Let's see if you have the idea. Classify each of the following elements.

a. $_{16}$S is a   _____ element.

b. $_{26}$Fe is a _R_____ element.

c. $_{19}$K is a _R_____ element.

d. $_{24}$Cr is a _____ element.

e. $_{18}$Ar is a _N_____ element.

f. $_{63}$Eu is a _R_____ element.

g. $_{92}$U is a _A_____ element.

1.3 Consider the following elements:

$$_{37}Rb \qquad _{35}Br \qquad _{95}Am$$
$$_{21}Sc \qquad _{54}Xe \qquad _{34}Se$$
$$_{12}Mg \qquad _{61}Pm$$
$$_{71}Lu \qquad _{42}Mo$$

Which of these are

a. representative elements _____

b. transition elements _____

c. actinides _____

d. noble gases _____

e. lanthanides _____

f. inner transition elements _____

g. rare earth elements _____

Electron Configurations.   The similarities in the properties of the elements can generally be attributed to either (a) a similarity in electron configuration or (b) a similarity in atomic radii or, more accurately, the charge densities (charge/radius). In this discussion we will only be concerned with the electron configuration.

Although mnemonic devices can be used to determine electron configurations, they should be avoided. Such devices merely give a means of obtaining a correct answer. They contribute nothing to our understanding of the basic principles. The purpose of learning electron configurations is to understand the periodic table. Consequently, the periodic table itself should be used for this purpose.

We should start by identifying the terms in an electron configuration. If we consider hydrogen, the lightest of all elements, the configuration is $1s^1$. These symbols represent the following:

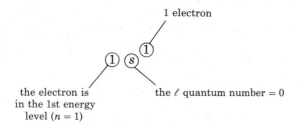

Now, by noting the general structure of the periodic table in terms of s, p, d, and f electrons, and by recognizing that the atomic number of an element is also the number of electrons in an atom of that element, it is possible to determine the electron configuration of any element. For instance, carbon is element number six, and it can be seen that it is the second element in the region where p orbitals are being filled, the first such element being boron.

Referring to Figure 1.3, we can start by noting that the first two electrons are accounted for by H and He, giving the configuration $1s^2$. This is

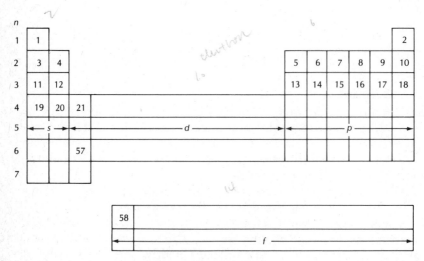

Figure 1.3   The periodic table with selected atomic numbers

true in spite of the fact that helium occurs with the noble gases.[1] The next two electrons, 3 and 4, are seen to fall into the region characterized by s electrons, giving $1s^2 2s^2$ for the first four electrons. Finally, the last two electrons, electrons 5 and 6, fall in the region characterized by p electrons, so the complete configuration for carbon is $1s^2 2s^2 2p^2$.

_____ **EXERCISE** _____

1.4 This should be easy. Using the periodic table as a guide, let's consider some electron configurations and the positions of the elements in the periodic table.

a. Sodium is the first element in the __3rd__ period filling in __s__ (type) electrons.

b. A fluorine atom contains ___9___ electrons, of which ___5___ are p electrons and ___4___ are s electrons.

c. The electron configuration of Ar is __$1s^2\ 2s^2\ 2p^6\ 3s^2\ 3p^6$__.

d. $1s^2 2s^2 2p^6 3s^2 3p^3$ is the electron configuration of ___P___ .

e. Element number 17 has the electron configuration __$1s^2\ 2s^2\ 2p^6\ 3s^2\ 3p^5$__.

_____

An important point can be made by considering the configuration of $_{19}$K and $_{21}$Sc. Potassium has an atomic number of 19 and, therefore, has 19 electrons. Proceeding through the periodic table in order of increasing atomic numbers, we obtain $1s^2$ for H and He, $1s^2 2s^2$ by adding Li and Be, and completing the second period to Ne, we have $1s^2 2s^2 2p^6$. The next element is Na which occurs in the next block of s electrons. Its configuration, then, is $1s^2 2s^2 2p^6 3s^1$, and Mg will complete the block of $3s$ electrons with the configuration $1s^2 2s^2 2p^6 3s^2$. The following six elements complete the p electrons in the third level, giving $1s^2 2s^2 2p^6 3s^2 3p^6$ for the configuration of Ar. The next element is K, element 19. One might expect it to be characterized by a $3d^1$ configuration since the d electrons first occur in the

_____

[1] Helium is different from the other noble gases in that it has no p electrons. It does, however, have a completely filled valence shell. In terms of electron configuration, it is awkward to place helium with the other noble gases, but it would certainly be wrong to place it at the top of family IIA on the basis of two s electrons. It is the filled valence shell that justifies helium's position among the noble gases.

third shell ($n = 3$). However, K falls in the block where s electrons are being filled in, thereby giving the configuration $1s^2 2s^2 2p^6 3s^2 3p^6 4s^1$.

Here we should recall that the elements in a given family are so arranged because of similarities in their properties. Thus, the properties of K are similar to those of the other members of family IA. And at the beginning we pointed out that this similarity is related to similarities in electron configurations.

It should be noted that the electron configuration of each member of family IA ends with ns^1; that is, there is always one electron in the s orbital of the outer shell. To emphasize this point, consider the configurations of the members of the representative element families shown below.

IA	IIIA
Li: $1s^2 2s^1$	B: $1s^2 2s^2 2p^1$
Na: $1s^2 2s^2 2p^6 3s^1$	Al: $1s^2 2s^2 2p^6 3s^2 3p^1$
K: ———————$4s^1$	Ga: ———————$4s^2 4p^1$
Rb: ———————$5s^1$	In: ———————$5s^2 5p^1$
Cs: ———————$6s^1$	Tl: ———————$6s^2 6p^1$

VIIA	
F: $1s^2 2s^2 2p^5$	
Cl: $1s^2 2s^2 2p^6 3s^2 3p^5$	
Br: ———————$4s^2 4p^5$	
I: ———————$5s^2 5p^5$	
At: ———————$6s^2 6p^5$	

Note that in each family the outer electron configurations are similar. It is this similarity that accounts for the similarity of chemical and physical properties within a family.

Now we can consider the configuration of $_{21}$Sc. It is seen to be the first element in a block of elements in which d electrons are being filled in, and since the d electrons first occur in the level $n = 3$, they are $3d$ electrons. Thus, the configuration for Sc is $1s^2 2s^2 2p^6 3s^2 3p^6 4s^2 3d^1$.

————————   **EXERCISE**   ————————

1.5 Using only the periodic table as a guide, write the electron configurations for

a. $_{14}$Si _____   **b.** $_{16}$S _____

c. $_{25}$Mn _____   **d.** $_{32}$Ge _____

e. $_{38}$Sr _____

====================================

It should be pointed out that minor variations sometimes occur among the electron configurations of the transition and the inner transition elements. These variations are not important for this discussion.

Shorthand Notation. It is rather time consuming to write out the electron configurations for the larger atoms. Thus, a shorthand notation is frequently used. You simply note the noble gas preceding the element in question, write the symbol for the noble gas in brackets, and then add the additional electrons. This notation can easily be understood if we consider the following:

He: $1s^2$ Li: $[1s^2]2s^1$ Li: $[He]2s^1$

Ne: $1s^2 2s^2 2p^6$ Na: $[1s^2 2s^2 2p^6]3s^1$ Na: $[Ne]3s^1$

Ar: $1s^2 2s^2 2p^6 3s^2 3p^6$ K: $[1s^2 2s^2 2p^6 3s^2 3p^6]4s^1$ K: $[Ar]4s^1$

Now let's consider $_{37}$Rb. The preceding noble gas is $_{36}$Kr. Thus, the electron configuration of Rb is

$$Rb: [Kr]5s^1$$

To take a more complicated case, consider $_{49}$In, with the configuration

$$In: [1s^2 2s^2 2p^6 3s^2 3p^6 4s^2 3d^{10} 4p^6]5s^2 4d^{10} 5p^1$$

Now note that the portion enclosed in brackets is the configuration of $_{36}$Kr. Thus, we can write the configuration for In as simply

$$In: [Kr]5s^2 4d^{10} 5p^1$$

Ionic Configurations. If we wish to write the configuration of an ion such as Li$^+$, we simply note that the ion has one less electron than the Li atom. Thus,

$$Li: 1s^2 2s^1 \quad \text{and} \quad Li^+: 1s^2 \quad \text{or} \quad [He]$$

For the representative elements, it is always the outer electrons that are lost.

Next, consider Al^{3+}. Here we have

$$Al: 1s^2 2s^2 2p^6 3s^2 3p^1 \quad \text{and} \quad Al^{3+}: 1s^2 2s^2 2p^6 \quad \text{or} \quad [Ne]$$

For a negative ion, one or more electrons are added to the atom. So, for F^-, we have

$$F: 1s^2 2s^2 2p^5 \quad \text{and} \quad F^-: 1s^2 2s^2 2p^6 \quad \text{or} \quad [Ne]$$

It is interesting to note that the electron configuration of Al^{3+} is the same as that of F^-.

EXERCISE

1.6 By now you should be pretty good at this. Give the electron configurations for

a. $_{12}Mg$ _____

b. $_{53}I$ _____

c. $_{30}Zn$ _____

d. $_{81}Tl$ _____

e. $_{71}Lu$ _____

f. $_{20}Ca^{2+}$ _____

g. $_{17}Cl^-$ _____

h. $_{16}S^{2-}$ _____

i. $_{21}Sc^{3+}$ _____

j. $_{19}K^+$ _____

Family Configurations. Recognizing family configurations is generally more useful than focusing on the configurations of individual elements. Thus, family IA is characterized by one electron in the s orbital of the outermost level, and the family configuration is ns^1, where n refers to the outer or valence shell. Each member of family IIA has two electrons in the s orbital of the outermost level and, therefore, the family configuration is ns^2. Continuing this process for the representative and noble gas elements gives rise to the family configurations shown in Figure 1.4.

For the transition elements, the family configurations must reflect the overlap of the orbitals. Thus, family IIIB (Sc, Y, La) can be characterized by the configuration $ns^2(n-1)d^1$. This is evident from the configuration we presented for Sc, $1s^2 2s^2 2p^6 3s^2 3p^6 \underline{4s^2 3d^1}$. Here we note that the d level is lower by one principal quantum number than the s level, a relationship that exists throughout the transition elements.

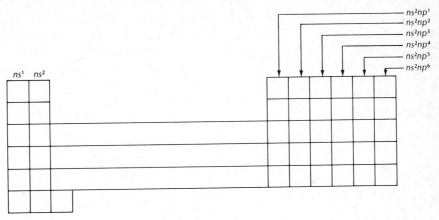

Figure 1.4   Family configurations for the representative elements

<hr>

EXERCISES
<hr>

1.7   Give the family electron configuration for

a. family IVA _____

b. F, Cl, Br, and I _____

c. Be, Mg, Ca, Sr, and Ba _____

d. family IIIB _____

e. the noble gases _____

1.8   a. Give the electron configurations for

   Zn _____

   Cd _____

   Hg _____

b. Zn, Cd, and Hg are in family _____ .

c. Zn, Cd, and Hg have the family configuration _____ .

<hr>

Orbital Configurations.   A useful method of expressing family electron configurations is in terms of the orbitals of the valence or outer shell. The electron distribution in the orbitals can be indicated with arrows. Two principles are used to determine the orbital distribution. The first principle states that only two electrons can exist in an orbital, and these must have opposite spins. The second principle, called Hund's rule, essentially states that the electrons go into a given set of orbitals unpaired until they are forced to pair up. Now let's apply these principles to family **IA**, where the configuration ns^1 becomes

$$\frac{\uparrow}{ns}$$

The arrow indicates that there is one electron in the outer s orbital. In a similar manner the family **IIA** configuration ns^2 becomes

$$\frac{\uparrow\downarrow}{ns}$$

and the family **IIIA** configuration ns^2np^1 becomes

$$\frac{\uparrow\downarrow}{ns} \quad \frac{\uparrow}{np}\,\frac{}{}\,\frac{}{}$$

In the next two examples, we see the application of Hund's rule. Here the family **VIA** configuration ns^2np^4 becomes

$$\frac{\uparrow\downarrow}{ns} \quad \frac{\uparrow\downarrow}{np}\,\frac{\uparrow}{}\,\frac{\uparrow}{}$$

and the family **VB** configuration $ns^2(n-1)d^3$ becomes

$$\frac{\uparrow\downarrow}{ns} \quad \frac{\uparrow}{(n-1)d}\,\frac{\uparrow}{}\,\frac{\uparrow}{}\,\frac{}{}\,\frac{}{} \quad \frac{}{np}\,\frac{}{}\,\frac{}{}$$

──────── **EXERCISE** ────────

1.9 Now let's see if you've got it. Using orbital notation, show the electron configuration for each of the following:

a. family **IVA** _____   **b.** family **VA** _____

c. family **VIIA** _____   **d.** family **IIB** _____

e. the noble gases _____

THE CHARGE ON AN ION

When a chemical compound is formed, we know that the atoms making up the compound are held together by some kind of chemical bond. The simplest of these is the ionic bond, which can be understood in terms of a positive charge attracting a negative charge. Thus, if we can understand why an ion should form and, further, what charge it should have, we should be able to construct some simple chemical compounds. This, we will see, is very easy to do.

In order for an atom to form a positive ion, it must lose one or more electrons. Only a few elements tend to form actual positive ions, but most atoms tend to form a species containing an apparent positive charge, and for the purpose of constructing compounds, the result is the same as it would be if these atoms were actually forming true ions. Thus, we shall see that this same method can be used to construct compounds that are predominantly covalent.

In order to form a negative ion, it will, of course, be necessary for an atom or group of atoms to gain one or more electrons. In the case of single atoms, this will be limited to the very electronegative elements.

Electronegativity.   It has long been recognized that in a chemical bond some atoms have a stronger attraction for electrons than do other atoms. For instance, in the HCl molecule, the Cl atom has a greater attraction for the electron pair in the bond than does the H atom. The term *electronegativity* is used to describe this effect. Thus, Cl is more electronegative than is H. The general trend of electronegativities is shown in Figure 1.5. Fluo-

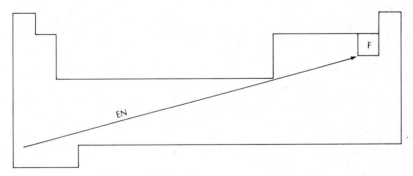

Figure 1.5   The periodic trend in electronegativity

rine is considered to be the most electronegative element. The first quanti-
tative measure of electronegativity was developed by Professor Linus
Pauling, and a table of his electronegativity values is given on the inside
back cover of your book.[2]

You will find electronegativities to be extremely easy to use and, in
addition, very important. Just to be sure that you understand the idea, try
these.

_____ EXERCISES _____

1.10 Use the general trend in electronegativities to determine which
element in each of the following pairs is more electronegative:

a. Na or Cl _____   **b.** Cl or Br _____

c. C or N _____   **d.** O or Cl _____

e. Al or F _____

1.11 Now use the Pauling table of electronegativity values to determine
which element in each pair is more electronegative:

a. C or H _____   **b.** P or I _____

c. Na or Cs _____   **d.** Si or I _____

e. Ge or Pb _____

Positive Ions. We have seen that the electron configuration for
members of family **IA** is ns^1. This means that for each member of the
family there is one electron in the s orbital of the outer or valence shell. In
a chemical reaction, this electron can be lost to form a 1+ ion. This is
readily understood from an examination of the ionization energies[3] of the

---

[2] Although there is a general understanding of the meaning of electronegativity, considerable
disagreement exists on how it should be measured. Consequently, there are many different
electronegativity scales. They generally agree, but there will be some differences. The electronegativity table at the back of this book uses values calculated by Allred based on the
Pauling approach. Do not be surprised if some of the values in this table differ from those in
other tables.

[3] The ionization energy (ionization potential) is the energy necessary to remove an electron
from a free gaseous atom or ion.

various electrons in a **IA** family member. For instance, let's look at the ionization energies for the three electrons in lithium.

Electron	I.E. (eV)
1	5.390
2	75.619
3	122.420

This table shows that the removal of the first electron requires a relatively small amount of energy, but it takes a considerable amount of energy to remove the next two electrons. Consequently, lithium tends to form an ion with a $1+$ charge but not $2+$ or $3+$. This same argument holds for all of the members of family **IA**.

As a further illustration of the principle, consider family **IIA**. Here the family configuration is ns^2, and using beryllium as the example, we find that the ionization energies for the four electrons in beryllium are as follows:

Electron	I.E. (eV)
1	9.320
2	18.206
3	153.850
4	217.657

Given these values, it is not surprising that members of family **IIA** form ions with a $2+$ charge.

Negative Ions.   Just as some atoms lose one or more electrons to become positive ions, other atoms can gain electrons to become negative ions. As an example, consider family **VIIA**, which is characterized by the configuration ns^2np^5. Using orbitals, the valence shell configuration can be represented as

$$\underset{ns}{\underline{\uparrow\downarrow}} \quad \underset{np}{\underline{\uparrow\downarrow}\ \underline{\uparrow\downarrow}\ \underline{\uparrow}}$$

This method of representation emphasizes the fact that one electron is missing in a **p** orbital. This position can be filled to give a completed valence shell and a resultant ion having a $1-$ charge. It is not likely that a member of this family would accept two electrons because the second electron would have to go into the $(n + 1)$ level, which is an excited state (a state of unstable energy).

In the same manner, family **VIA** is characterized by the configuration

ns^2np^4, and in terms of the orbital population, the distribution of electrons is

$$\underset{ns}{\underline{\uparrow\downarrow}} \quad \underset{np}{\underline{\uparrow\downarrow \ \uparrow \ \uparrow}}$$

We would expect members of this family to accept two electrons to give a $2-$ charge. You might note that in the formation of a negative ion, the number of electrons gained is the number necessary to attain a noble gas configuration.

The very electronegative elements, such as those in families VIIA and VIA along with N and P, will gain electrons to form negative ions, whereas the metals will lose electrons to form positive ions.

—————— **EXERCISES** ——————

1.12 Fill in the following blanks by relating electron configuration to ionic charge.

a. $_{16}$S is in family _____, which has the family electron

configuration _____, and it should form a negative ion

with a charge of _____.

b. $_{55}$Cs is in family _____, which has the family electron

configuration _____, and it should form a positive ion

with a charge of _____.

c. $_{35}$Br is in family _____, which has the family electron

configuration _____, and it should form a negative ion

with a charge of _____.

1.13 What charge would you expect for the ion formed from each of the following atoms?

a. $_9$F _____   **b.** $_8$O _____

c. $_7$N _____   **d.** $_{19}$K _____

e. $_{17}$Cl _____   **f.** $_{16}$S _____

g. $_{56}$Ba _____   **h.** $_{53}$I _____

i. $_{37}$Rb _____   **j.** $_{34}$Se _____

IONIC COMPOUNDS

When an ionic compound is formed, it is necessary that the charges on the ions be neutralized. Thus, for every positive charge there must be an equal negative charge to cancel it. For example, the 1+ charge of a Na^+ ion may be neutralized by the 1− charge of a Cl^- ion, or the 2+ charge of a Ca^{2+} ion may be neutralized by two Cl^- ions, as we see in the reactions

$$Na^+ + Cl^- \longrightarrow NaCl$$
$$Ca^{2+} + 2\,Cl^- \longrightarrow CaCl_2$$

For the balanced chemical reaction showing the formation of NaCl, we would write

$$2\,Na + Cl_2 \longrightarrow 2\,NaCl$$

In this instance, we are saying that sodium reacts with Cl_2 to form NaCl. And, although the element chlorine consists of Cl_2 molecules, this does not affect the formula of the compound NaCl. Sodium chloride has one Na and one Cl because Na forms a 1+ ion (family **IA**) and Cl forms a 1− ion (family **VIIA**).

For the reaction between Ca and Cl_2 to form $CaCl_2$, we would write

$$Ca + Cl_2 \longrightarrow CaCl_2$$

As a third example, imagine that you are given the reactants Mg and O_2 and are asked to predict the product of the reaction. First you should note that magnesium is in family **IIA**, and therefore an atom of magnesium tends to lose 2 electrons to become a 2+ ion. Next you should note that oxygen is in family **VIA**, and is the more electronegative of the two atoms in question. Thus, oxygen will tend to form a negative ion. Its orbital configuration is

$$\underset{1s}{\underline{\uparrow\downarrow}} \quad \underset{2s}{\underline{\uparrow\downarrow}} \quad \underset{2p}{\underline{\uparrow\downarrow \,\, \uparrow \,\, \uparrow}}$$

and it can accept 2 electrons to give a completed *p* level and, consequently, a 2− ion. Thus, oxygen should form the ion O^{2-}. It is then easy to see that the formula of a compound between magnesium and oxygen will have to be MgO. Thus the reaction is

$$Mg + O_2 \longrightarrow MgO$$

and when it is balanced we obtain

$$2\,Mg + O_2 \longrightarrow 2\,MgO$$

Note that the positive species has been written first in these compounds. There are a few exceptions to this rule, but this is the accepted convention.

───────── **EXERCISE** ─────────

1.14   Here we see the first application of the principles we've learned. What would you expect for the formula of a compound formed between

a. Cl and Mg _____   **b.** Na and S _____

c. Ca and F _____   **d.** Sr and I _____

e. O and K _____   **f.** Se and Rb _____

g. P and Ba _____   **h.** Ca and S _____

i. Na and N _____   **j.** Te and Cs _____

═══════════════════

OXIDATION STATE AND VALENCE

Although they are totally different quantities, valence and oxidation state are frequently confused. The valence of an atom may be considered to be its combining capacity as measured by the number of single bonds it forms. Thus, for water,

$$H-O_{\diagdown H}$$

oxygen has a valence of two whereas hydrogen has a valence of one. Note that this is not 2− or 1+, but simply 2 and 1, since combining capacity is always a positive or zero quantity. In order to determine the valence of an atom, it is necessary to know the structure of the compound in which the atom is located. A much more useful concept than valence is that of *oxidation state* (or *oxidation number*). This is *the apparent charge on an atom based upon a set of rules*. In general, these rules are quite reasonable, and you have probably been using some of them without realizing their relation to the oxidation states of the elements. For instance,

1. The oxidation state of a free element is zero.
2. The oxidation state of a monatomic ion is the charge on the ion. Thus, the oxidation state of Ca^{2+} is 2+ and that of Cl^- is 1−.

3. In covalent compounds, the oxidation state of an element is determined by assigning each shared electron pair to the more electronegative atom. In the case of water,

$$H \overset{\circ}{\underset{\circ}{\cdot}} O \overset{\circ}{\underset{\circ}{\cdot}} H$$

both electron pairs are assigned to the oxygen atom. This assignment results in the presence of two extra electrons around the oxygen atom, thereby giving it an oxidation state of $2-$, and each hydrogen atom will have lost one electron, thus giving each hydrogen atom an oxidation state of $1+$.

_____ **EXERCISES** _____

1.15 In these, let's see if you now know the difference between valence and oxidation state.

	Valence	Oxidation state
a. S in H_2S	_____	_____
b. Cl in HCl	_____	_____
c. H in H_2O	_____	_____
d. C in CCl_4	_____	_____
e. N in NH_3 [4]	_____	_____

1.16 And what is the oxidation state of

a. Se in Rb_2Se _____

b. I in CaI_2 _____

c. Cs in Cs_2O _____

d. Na in Na_3N _____

e. S in SrS _____

_____

[4] NH_3 is an exception to the rule that the positive species is written first in a compound.

PREDICTING OXIDATION STATES

When we predicted the charges on ions based on electron configurations, we were, in fact, predicting oxidation states, because the charge on a monatomic ion is its oxidation state. But interestingly, we can use the same principle to predict the oxidation states of atoms in covalent molecules. Actually, that is what we just did above when we used H_2O as an example.

If we focus on the representative elements, we find that the most common positive oxidation states can be predicted by imagining the loss of electrons in the manner illustrated earlier for positive ions. Although ions are not actually formed in covalent compounds, the electrons will be preferentially attracted to the more electronegative element. Therefore, we can assume that the more electropositive element tends to give up its electrons, and the number of electrons lost will be determined by the electron configuration.

We have already seen that the elements in family IA (ns^1) lose one electron to form $1+$ ions and members of family IIA (ns^2) lose two electrons to form $2+$ ions. The remainder of the representative elements have both s and p electrons. There is enough energy difference between these levels that the p electrons can be donated without the loss of the s electrons. If we consider family IIIA, the family electron configuration is seen to be ns^2np^1 or, in terms of the orbital electron distribution,

$$\underset{ns}{\underline{\downarrow\uparrow}} \quad \underset{np}{\underline{\uparrow}\ \underline{}\ \underline{}}$$

The donation of only the p electron would result in a $1+$ oxidation state, and $1+$ is, in fact, a commonly observed oxidation state for all of the members of this family except boron and aluminum. But it is also possible for all three of the electrons in the valence shell to be involved, in which case the oxidation state is $3+$. For family IIIA, there are, thus, two common oxidation states observed, and these are $1+$ and $3+$ as predicted from the electron configurations.

Next consider family VA with the family electron configuration

$$\underset{ns}{\underline{\downarrow\uparrow}} \quad \underset{np}{\underline{\uparrow}\ \underline{\uparrow}\ \underline{\uparrow}}$$

Here the tendency to lose the three p electrons leads us to expect a $3+$ oxidation state, and a $5+$ oxidation state should occur if the s electrons are also involved. For the members of this family (N, P, As, Sb, and Bi), $3+$

and 5+ are the most common positive oxidation states. This same argument can be applied to the rest of the representative element families.

Just as it is possible to give up electrons to form positive oxidation states, electrons can be gained to form negative oxidation states. Further, this can be treated, in general, in the same manner as was done earlier for negative ions. Consider, for instance, family VIA, which has the configuration

$$\underset{ns}{\uparrow\downarrow}\quad \underset{np}{\uparrow\downarrow\ \uparrow\ \uparrow}$$

There are two orbitals that can each accept one electron. Thus, members of this family can gain two electrons and exhibit an oxidation state of 2−. For family VA, we can see from the orbital configuration on p. 20 that members of this family can accept three electrons to give a filled valence shell (a noble gas configuration). Consequently, we would expect members of this family to have a negative oxidation state of 3−. As it turns out, only families IVA to VIIA tend to show negative oxidation states, with the tendency increasing markedly toward family VIIA.

In summary, we find the following oxidation states for the representative elements:

Family	Family configuration	Predicted oxidation states
IA	ns^1	1+
IIA	ns^2	2+
IIIA	ns^2np^1	1+, 3+
IVA	ns^2np^2	2+, 4+ 4−
VA	ns^2np^3	3+, 5+ 3−
VIA	ns^2np^4	4+, 6+ 2−
VIIA	ns^2np^5	5+, 7+ 1−

_____ **EXERCISES** _____

Let's review family electron configurations and then apply them to the prediction of oxidation states.

1.17   Give the family orbital configuration for

a. $_{20}$Ca _____   **b.** $_{16}$S _____

c. $_{17}$Cl _____   **d.** $_{31}$Ga _____

e. $_{11}$Na _____   **f.** $_{81}$Tl _____

g. $_{33}$As _____   **h.** $_{50}$Sn _____

i. $_{56}$Ba _____   **j.** $_{34}$Se _____

1.18   List all of the oxidation states you would predict for

a. $_{20}$Ca _____   **b.** $_{16}$S _____

c. $_{17}$Cl _____   **d.** $_{31}$Ga _____

e. $_{11}$Na _____   **f.** $_{49}$In _____

g. $_{83}$Bi _____   **h.** $_{32}$Ge _____

i. $_{38}$Sr _____   **j.** $_{52}$Te _____

Formulas of Compounds Based on Oxidation States.   Once you know the oxidation states of the elements, it is very easy to figure out the formulas of the simpler compounds. To begin with, let's consider compounds containing only two elements. When two different elements combine to form a compound, one of the elements will be more electronegative than the other. Consequently, that element will attract the electrons more strongly toward itself, resulting in a negative oxidation state. The other element will tend to give up its electrons, thereby resulting in a positive oxidation state. The determination of the positive and negative species will, of course, require that we recall the general trend in electronegativity values (see Figure 1.5).

Now let's consider a compound formed between thallium (element 81) and oxygen. From the general trend in electronegativity, we readily conclude that oxygen is the more electronegative of the two elements and, therefore, will exhibit the negative oxidation state. Noting that oxygen is in family **VIA** with the configuration

$$\underset{ns}{\underline{\uparrow\downarrow}} \quad \underset{np}{\underline{\uparrow\downarrow \;\; \uparrow \;\; \uparrow}}$$

it is easy to see that it will accept two electrons to give a 2− oxidation state. Thallium has the family configuration ns^2np^1 and can lose the **p** electron to give a 1+ oxidation state or lose the two **s** electrons along with the **p** electron to give a 3+ oxidation state. Thus, we have

$$Tl^{1+} \qquad O^{2-}$$

$$Tl^{3+}$$

There are two possible ways of combining these elements to obtain a charge balance:

$$\overset{1+\ 2-}{Tl_2O} \qquad \text{and} \qquad \overset{3+\ 2-}{Tl_2O_3}$$

Tl: $2 \times 1+ = 2+$	Tl: $2 \times 3+ = 6+$
O: $1 \times 2- = 2-$	O: $3 \times 2- = 6-$
0	0

Actually, there are two compounds formed between thallium and oxygen, and their formulas are those we just predicted. Certainly, it is impressive to be able to predict the formulas of the compounds formed between two elements even though you have never seen those compounds or, probably, never even heard of them.

Let's consider one more example, in which a compound is formed between phosphorus and chlorine. Chlorine is more electronegative than phosphorus and will thus have a negative oxidation state. Phosphorus, then, must have a positive oxidation state. Based on the electron configurations, we can say the following:

Element	Family configuration	Predicted oxidation states	Oxidation states in this instance
P	ns^2np^3	$3+, 5+, 3-$	$3+, 5+$
Cl	ns^2np^5	$5+, 7+, 1-$	$1-$

From this information, we can conclude that two compounds can be formed between P and Cl with the formulas

$$\overset{3+}{P\ Cl_3} \qquad \text{and} \qquad \overset{5+}{P\ Cl_5}$$

P: $1 \times 3+ = 3+$	P: $1 \times 5+ = 5+$
Cl: $3 \times 1- = 3-$	Cl: $5 \times 1- = 5-$
0	0

_____ **EXERCISE** _____

1.19 This is the next major step. See if you can give the formula(s) of all compounds you would predict to be formed between

a. Al and S _____   **b.** N and F _____

c. O and Ge _____   **d.** Cl and Te _____

e. In and Cl _____   **f.** Sn and Se _____

g. O and Pb _____   **h.** N and Li _____

i. Tl and O _____   **j.** I and F _____

_____

Oxidation States of the Transition Elements.   The oxidation states of the transition elements are not arrived at in so straightforward a manner as are those of the representative elements. In some cases it is necessary to memorize the oxidation states, but generally they follow a few simple rules. So for the transition elements we can say

1. They will have positive oxidation states (all transition elements are metals).
2. Oxidation states of $2+$ and $3+$ are very common.
3. Multiple oxidation states are usual. It would be rather difficult to learn all of the values, so you should concentrate on only the most common ones.
4. Elements in families IIIB through VIIB commonly have an oxidation state equal to the group number. This means that both the outermost *s* and *d* electrons are involved.
5. The most common values for the more familiar group VIII elements are as follows:

$$\begin{array}{ll} \text{Fe:} & 3+, 2+ \\ \text{Co:} & 3+, 2+ \\ \text{Ni:} & 2+ \\ \text{Pt:} & 4+, 2+ \end{array}$$

6. For family **IB**, the most common values are

$$\begin{array}{ll} \text{Cu:} & 2+, 1+ \\ \text{Ag:} & 1+ \\ \text{Au:} & 3+, 1+ \end{array}$$

7. In family **IIB**, since the *d* orbitals are completely filled and only the *s* electrons can be involved, we would expect only a value of $2+$, corresponding to the involvement of the two *s* electrons.

The most common oxidation states of a selected group of the transition elements are summarized in Table 1.1.

Table 1.1 Common Oxidation States of Selected Transition Elements

Sc	Ti	V	Cr	Mn	Fe	Co	Ni	Cu	Zn
3+	4+	5+	6+	7+	3+	3+	2+	2+	2+
	3+	4+	3+	4+	2+	2+		1+	
		3+		3+					
				2+					
Y	Zr	Nb	Mo	—	—	—	—	Ag	Cd
3+	4+	5+	6+					1+	2+
La	Hf	Ta	W	—	—	—	Pt	Au	Hg
3+	4+	5+	6+				4+	3+	2+
							2+	1+	1+

The Inner Transition Elements. Of the inner transition series, we will consider only the lanthanides, elements 58 through 71. The most common, and almost the only, oxidation state for all of these elements is $3+$, which results from the involvement of the $6s^2 5d^1$ electrons.

As an example, consider the compound formed between praseodymium (element 59) and sulfur. All of the inner transition elements are metals and show only positive oxidation states. Sulfur, then, is the more electronegative of the two elements and will have an oxidation state of $2-$. Praseodymium, being a lanthanide, will have an oxidation state of $3+$. Thus, the compound will be

$$\overset{3+\;\;2+}{Pr_2 S_3}$$

Pr: $2 \times 3+ = 6+$
S: $3 \times 2- = \underline{6-}$
$$0$$

Actually, most of the elements can have oxidation states other than those shown. However, we have presented here the most commonly observed states.

_____ **EXERCISE** _____

1.20 Now we can go beyond the representative elements. Based on the rules you have learned, and the table of oxidation states where necessary, give the formula(s) of the compounds you would expect to be formed between

a. O and Zn _____   **b.** Cl and Ni _____

c. Nb and S _____   **d.** Se and Nd _____

e. Cu and I _____   **f.** Sc and O _____

g. N and Tm _____   **h.** Fe and F _____

i. Zr and S _____   **j.** Mo and O _____

VALENCE AND OXIDATION STATE REVISITED

We have already noted that valence and oxidation state often have the same numerical value. Using water as our example,

$$H-O_{\diagdown_{\textstyle H}}$$

we saw that oxygen forms two single bonds and, therefore, has a valence of 2, whereas each hydrogen atom forms one single bond and has a valence of 1. From our earlier discussion, we also noted that oxygen has an oxidation state of $2-$ and hydrogen has an oxidation state of $1+$. Thus, the numerical values of the oxidation states and the valences are the same.

If we now consider hydrogen peroxide, H_2O_2, we find a quite different situation. The structure of H_2O_2 is

$$H-O-O-H$$

and we see that since each oxygen atom forms two single bonds, oxygen again has a valence of 2. Similarly, the valence of hydrogen is again 1. However, from the formula H_2O_2, we can see that the oxidation states cannot be $2-$ and $1+$. In order to have a balance of charge, if hydrogen is $1+$, then oxygen must be $1-$. On the other hand, if we let oxygen have an oxidation state of $2-$, then hydrogen must have a value of $2+$. Either of these sets of values might be used, but it is more reasonable to let H $= 1+$ and O $= 1-$. The reason for the different oxidation state for oxygen in this instance is that the oxygen atoms are bonded to each other. Anytime a compound contains atoms of the same element that are bonded to each other, there will be a difference between the numerical values of the valence and the oxidation state for that element in that particular compound. Sometimes this difference leads to some rather strange results, as we will see in the next exercise. It also explains why mercury can show an oxidation state of 1+ in addition to its expected oxidation state of 2+ (see part d of Exercise 1.21).

───────────── **EXERCISE** ─────────────

1.21 Draw the structure of the compound and give the valence and oxidation state of the underlined element for

	Structure	Valence	Oxidation state
a. $\underline{C}H_4$	H | H—C—H | H	4	4−
b. $\underline{C}_2H_6$			
c. $\underline{C}_3H_8$			
d. $\underline{Hg}_2Cl_2$			
e. $\underline{N}_2H_4$			

A PREVIEW OF DIRECT COMBINATION REACTIONS

Remember, that the purpose of this book is to help you to predict the products of chemical reactions. And although you may not have realized it, all that you have learned in Chapter 1 has been directed toward this goal. Now you should be ready to put this knowledge to work. But first, we want to point out that sometimes products other than those predicted will result, and sometimes what you predict will be wrong. But we also want to point out that even though your prediction may turn out to be wrong, it will be reasonable. It should also be recognized that many binary compounds cannot be made by simply combining the two elements. At this point, it is not possible for you to know which products are of this type.

The important thing is that now that you know how to write the formulas of binary compounds, it is possible for you to make reasonable predictions of the products of direct combination reactions.

Let's take one final example to illustrate how easy it is. What product would you expect from the reaction

$$Sb + F_2 \longrightarrow$$

First you note that fluorine is the more electronegative of the two and will, therefore, have a negative oxidation state. This state, of course, must be $1-$. Antimony would be expected to show two positive oxidation states, $3+$ and $5+$. Thus, two compounds can be formed. The formulas of these compounds are determined by balancing the oxidation states to give a net charge of zero.

$$Sb + F_2 \longrightarrow SbF_3 \quad \text{and} \quad Sb + F_2 \longrightarrow SbF_5$$

If we balance the equations, we obtain

$$2\ Sb + 3\ F_2 \longrightarrow 2\ SbF_3 \quad \text{and} \quad 2\ Sb + 5\ F_2 \longrightarrow 2\ SbF_5$$

In the next exercise, the last of this chapter, you now have a chance to prove that you have understood those topics we've discussed. If you can answer all of the parts of Exercise 1.22, you are ready to go on to the next chapter.

————————   **EXERCISE**   ————————

1.22   Complete the following reactions. If more than one product might be formed, show all the reasonable products. It is not necessary to balance the equations.

a. $Na + Cl_2 \longrightarrow$ _____   **b.** $C + O_2 \longrightarrow$ _____

c. $Ca + O_2 \longrightarrow$ _____   **d.** $Ge + Cl_2 \longrightarrow$ _____

e. $Te + O_2 \longrightarrow$ _____   **f.** $Li + N_2 \longrightarrow$ _____

g. $Pb + F_2 \longrightarrow$ _____   **h.** $In + I_2 \longrightarrow$ _____

i. $Sb + O_2 \longrightarrow$ _____   **j.** $Fe + Cl_2 \longrightarrow$ _____

k. $Tb + S_8 \longrightarrow$ _____   **l.** $Ni + F_2 \longrightarrow$ _____

m. $Cu + O_2 \longrightarrow$ _____   **n.** $Lu + Cl_2 \longrightarrow$ _____

o. $Ti + F_2 \longrightarrow$ _____   **p.** $Zn + O_2 \longrightarrow$ _____

————————————————————

2

Direct Combination Reactions

Based on the oxidation states of the combining elements, you learned in Chapter 1 how to predict the formulas of *binary* compounds (compounds containing only two elements). This, however, does not tell you which elements will, in fact, react by direct combination. Some will and some will not. Consequently, you can write down the products of a reaction without concern as to whether the reaction actually does take place. Or you can go to the trouble of memorizing all of the direct combination reactions. Neither of these approaches is desirable but, in fact, the first is closer to the intent of this book. In this chapter, we will continue to emphasize the use of oxidation states as a basis for predicting the products of direct combination reactions, but we will also apply these principles to reactions that actually do take place. It is, therefore, hoped that the student will end up somewhere between these two extremes.

Before considering any specific reactions, there are a few general rules that will prove helpful:

1. Among the representative elements, when two or more positive oxidation states are possible, the higher one is favored by the lighter members of the family. Thus, in family IIIA, where 3+ and 1+ are expected, 3+ is the more common oxidation state for Al whereas 1+ is the more common oxidation state for Tl. In family IVA, where 2+ and 4+ are the expected positive states, the 2+ state is by far the more common for Pb.

2. Among the transition elements, multiple oxidation states are the rule, but the higher state is favored by the heavier member of a family. Thus, we find that Cr^{3+} is more common than Cr^{6+}, whereas for tungsten W^{6+} is the most common oxidation state.

3. These distinctions are blurred for elements in the middle of a family. Additionally, the effect is not quite so pronounced among the transition elements as it is among the representative elements. It should also be noted that we see metals react with nonmetals and nonmetals react with other nonmetals, but we do not normally see metals react with other metals by direct combination.

_____   **EXERCISE**   _____

2.1 Now let's try a little review. For the following elements, the most common positive oxidation state is

a. Pr _____   **b.** B _____

c. Ta _____   **d.** Sn _____

e. Bi _____   **f.** Cr _____

g. Al _____   **h.** W _____

i. Sr _____   **j.** Tl _____

Using the principles developed in Chapter 1, it is a simple matter to predict reasonable products of direct combination reactions. For instance, the compound formed by the combination of sodium and chlorine is NaCl. This conclusion is based on the 1+ oxidation state of Na and the 1− oxidation state of Cl, both of which can be determined by considering the electronegativities and the electron configurations of the two elements. If this reaction is expressed as a balanced chemical equation, we have

$$2\,Na + Cl_2 \longrightarrow 2\,NaCl$$

The sodium is written as a monatomic species because it does not exist as a distinct molecule, whereas chlorine is written as Cl_2 because it occurs in nature as a diatomic molecule. But neither of these, the monatomic nature of sodium or the diatomic structure of chlorine, has anything to do with the formula of the product, NaCl.

To further emphasize this point, consider the reaction between bismuth and chlorine,

$$2 \text{ Bi} + 3 \text{ Cl}_2 \longrightarrow 2 \text{ BiCl}_3$$

Here the product $BiCl_3$ is determined from the expected oxidation states of Bi $(3+)$ and chlorine $(1-)$.

_____ **EXERCISE** _____

2.2 Now let's see if you can use what you learned in Chapter 1. Complete and balance the following reactions based on your predictions of the oxidation states.

a. $Ba + Br_2 \longrightarrow$ _____   **b.** $K + I_2 \longrightarrow$ _____

c. $Ca + Cl_2 \longrightarrow$ _____   **d.** $Gd + S_8 \longrightarrow$ _____

e. $Cs + Br_2 \longrightarrow$ _____   **f.** $Sc + F_2 \longrightarrow$ _____

g. $Mg + O_2 \longrightarrow$ _____   **h.** $La + F_2 \longrightarrow$ _____

i. $Zn + O_2 \longrightarrow$ _____   **j.** $Dy + Br_2 \longrightarrow$ _____

REACTIONS WITH OXYGEN

Normally, oxygen reacts in such a manner that the oxide ion (O^{2-}) is formed. However, there are two other oxygen anions that are sometimes observed. These are the peroxide anion (O_2^{2-}) and the superoxide anion (O_2^-). In order to know which one of the oxygen anions you have, it is only necessary to determine the oxidation states.

In the compound

$$TeO_2$$

Te will have either a $4+$ or a $6+$ oxidation state. If the oxidation state is $4+$, then we have O_2^{4-}, or O^{2-} for each oxygen, which corresponds to the oxide ion. If Te has an oxidation state of $6+$, we would have O_2^{6-}, where each oxygen atom has an oxidation state of $3-$. This doesn't correspond to anything you or I have ever seen. Thus, we conclude that, in this case, Te has an oxidation state of $4+$, and the anion is the oxide ion (O^{2-}).

Let's now consider

$$BaO_2$$

We know that Ba shows only a $2+$ oxidation state. Consequently, the anion must be O_2^{2-}, which is the peroxide ion. Finally, considering

$$KO_2$$

we easily conclude that since the oxidation state of potassium is $1+$, the anion is O_2^-, which corresponds to the superoxide ion.

To make it even easier, except for hydrogen peroxide, most of the peroxides and superoxides you are likely to see occur with members of families IA and IIA.

_____ **EXERCISE** _____

2.3 Identify each of the following compounds as to whether it is an oxide, a peroxide, or a superoxide.

a. Na_2O_2 _____   **b.** CrO_3 _____

c. H_2O_2 _____   **d.** BaO_2 _____

e. K_2O _____   **f.** SO_2 _____

g. CsO_2 _____   **h.** SrO _____

i. KO_2 _____   **j.** SO_3 _____

Reactions of O_2 with Metals. Virtually all of the metals react with O_2 by direct combination to form the oxide. In fact, it is much easier to list the elements that do not react with O_2 by direct combination than to list those that do. However, the normal product of the direct combination reaction of O_2 with an alkali metal, as well as with barium, is not the oxide. Sodium and barium react with O_2 to form the peroxide, whereas potassium, rubidium, and cesium react with O_2 to form the superoxide. All of the remaining elements react with O_2 to form the normal oxide.

_____ **EXERCISE** _____

2.4 Complete and balance the following reactions, being careful to distinguish between the oxides, superoxides, and peroxides.

a. $Ca + O_2 \longrightarrow$ _____   **b.** $Ba + O_2 \longrightarrow$ _____

c. $Na + O_2 \longrightarrow$ _____   **d.** $Mg + O_2 \longrightarrow$ _____

e. $Li + O_2 \longrightarrow$ _____   **f.** $K + O_2 \longrightarrow$ _____

g. $Rb + O_2 \longrightarrow$ _____   **h.** $Sr + O_2 \longrightarrow$ _____

i. $H_2 + O_2 \longrightarrow$ _____   **j.** $Cs + O_2 \longrightarrow$ _____

Metals with More Than One Oxidation State. When a metal can have two or more positive oxidation states, then two or more binary compounds are possible. In fact, several different binary compounds may exist for a given metal. The particular one that results from a direct combination reaction depends on such factors as temperature, pressure, and concentrations. One of these will usually be much more stable than the others, but it is not always obvious which one this will be. Consequently, even if you should be able to make a reasonable prediction based on the rules you have learned, it should be realized that the proposed reaction may not, in fact, occur. To illustrate the point, consider the chlorides of family IIIA. You learned in Chapter 1 that these show oxidation states of $1+$ and $3+$. Thus, for aluminum one should expect

$$2\,Al + Cl_2 \xrightarrow{\;\triangle\;} 2\,AlCl$$

and

$$2\,Al + 3\,Cl_2 \longrightarrow 2\,AlCl_3$$

You have also just learned that the $3+$ oxidation state of aluminum is the more common, and therefore you would expect $AlCl_3$ to be favored. Actually, both $AlCl$ and $AlCl_3$ can be made, but $AlCl$ exists only at very high temperatures. Interestingly, use of the $1+$ oxidation state has been proposed for the commercial production of aluminum, particularly for the recovery of aluminum from salvage such as aluminum cans. The electrolytic process presently used is highly energy intensive and, with the present cost of energy, it is correspondingly very expensive. Basically, the procedure involves the formation of $AlCl$ at high temperatures. It is then permitted to cool down to lower temperatures where it disproportionates. The equations are

$$2\,Al + Cl_2 \xrightarrow{\;\triangle\;} 2\,AlCl$$

$$3\,AlCl \longrightarrow 2\,Al + AlCl_3$$

The pure aluminum is removed and more scrap aluminum added. This is heated with the $AlCl_3$ to a high enough temperature to form $AlCl$, which again disproportionates when cooled. With variations, the procedure can also be used with aluminum ores.

Returning to the question of which chloride of aluminum will be formed, you should have expected $AlCl_3$, and you would be correct. Likewise, following the rule you learned at the beginning of this chapter, you would expect TlCl to be more stable than $TlCl_3$. However, when we consider indium, the decision between InCl and $InCl_3$ is not at all clear, because indium is a middle element in family IIIA. We have used the chlorides here as convenient examples, but the same arguments, of course, will also apply to other binary compounds of family IIIA. The oxides are not as straightforward as the chlorides, but for aluminum essentially only the 3+ state is observed, and the oxide will be Al_2O_3.[1]

<hr>

EXERCISE

2.5 Based on common oxidation states, give the expected products of each of the following reactions, and underline the one you would expect to be the most stable.

a. $Cr + O_2 \longrightarrow$ _____   **b.** $Al + O_2 \longrightarrow$ _____

c. $Bi + O_2 \longrightarrow$ _____   **d.** $Sn + O_2 \longrightarrow$ _____

e. $Tl + O_2 \longrightarrow$ _____   **f.** $W + O_2 \longrightarrow$ _____

<hr>

Reactions with Nonmetals. Oxygen does not react by direct combination with the halogens (family VIIA). However, oxygen does react with the members of families IVA, VA, and VIA and, except for nitrogen, gives the expected compounds based on oxidation states. There are six oxides of nitrogen, but only NO, NO_2, and N_2O_4 result from direct combination reactions. From the table of electronegativities, it can be seen that oxygen is the second most electronegative element. Consequently, it will show a negative oxidation state in compounds with all of the nonmetals except fluorine.

<hr>

[1] A naturally occurring form of Al_2O_3 is known by the mineral name, *corundum*. It is the second hardest of all minerals after diamond. Interestingly, two of the well-known gems, sapphire and ruby, are corundum crystals that are contaminated with trace impurities. Ruby contains traces of Cr_2O_3, and in sapphires the contaminants are FeO and TiO_2.

———————— **EXERCISE** ————————

2.6   As in the previous exercise, give the products you would expect based on oxidation states. Again, if two oxides are possible, underline the one you would expect to be the more stable.

a. $C + O_2 \longrightarrow$ _____   **b.** $Sb + O_2 \longrightarrow$ _____

c. $Ge + O_2 \longrightarrow$ _____   **d.** $Se + O_2 \longrightarrow$ _____

e. $P_4 + O_2 \longrightarrow$ _____   **f.** $S_8 + O_2 \longrightarrow$ _____

g. $Te + O_2 \longrightarrow$ _____   **h.** $As + O_2 \longrightarrow$ _____

REACTIONS WITH HYDROGEN

The one direct combination reaction of H_2 that is well known to every chemistry student is

$$2\,H_2 + O_2 \longrightarrow 2\,H_2O$$

and here we observe the expected product based on the oxidation states; namely, $H = 1+$ and $O = 2-$.

There are two additional things that are important to know about hydrogen. First, H_2 does not react with every element by direct combination. For instance, since Te is in the same family as oxygen, one would expect to find that

$$2\,H_2 + Te \longrightarrow 2\,H_2Te$$

Hydrogen telluride can be made but not by a direct combination reaction. Similar behavior is also observed between hydrogen and a large number of both metals and nonmetals. Second, hydrogen can show a $1-$ as well as a $1+$ oxidation state. With the very electropositive elements, hydrogen forms *saline* or *salt-like* hydrides. These are ionic compounds and are often referred to as *ionic* hydrides. As it turns out, there are very few direct combination reactions with hydrogen that need be considered.

———————— **EXERCISE** ————————

2.7   Before going on, be sure you can determine the oxidation state of hydrogen in each of the following. Remember to consider the electronegativity of the element bonded to hydrogen. If necessary, use the electronegativity table.

a. CaH_2 _____   b. SbH_3 _____

c. NH_3 _____   d. LiH _____

e. H_2Se _____   f. H_2S _____

g. CH_4 _____   h. HBr _____

i. CsH _____   j. SnH_4 _____

_____

Saline Hydrides. The saline hydrides are formed by the direct combination of hydrogen with elements of families **IA** and **IIA** (except Be) at temperatures between 300° and 700° C. Magnesium additionally requires a high H_2 pressure. The products of these reactions are just what you should expect from the oxidation states. Thus, the prediction of the correct products is simple once it is realized that this type of reaction occurs.

_____ **EXERCISE** _____

2.8   Try these. Complete the following equations, being careful to note all oxidation states.

a. $Li + H_2 \longrightarrow$ _____   b. $Ba + H_2 \longrightarrow$ _____

c. $Rb + H_2 \longrightarrow$ _____   d. $Ca + H_2 \longrightarrow$ _____

e. $Na + H_2 \longrightarrow$ _____   f. $Cs + H_2 \longrightarrow$ _____

_____

Nonmetal Hydrides.   Although hydrides can be made with essentially all of the representative elements, the elements of families **IIIA** and **IVA** do not react with hydrogen by direct combination. In families **VA** and **VIA**, only N_2 and O_2 react with hydrogen by direct combination, but in family **VIIA**, all of the members react with hydrogen. The nonmetal hydrides are easily remembered if the nonmetals are shown as they occur in the periodic table:

$$
\begin{array}{ccc}
N & O & F \\
  &   & Cl \\
  &   & Br \\
  &   & I
\end{array}
$$

_____   **EXERCISE**   _____

2.9 Now try some more. Complete the following equations, again being careful to note all oxidation states. If no reaction occurs, just write NR.

a. $Sn + H_2 \longrightarrow$ _____

b. $H_2 + I_2 \longrightarrow$ _____

c. $Cl_2 + H_2 \longrightarrow$ _____

d. $Sb + H_2 \longrightarrow$ _____

e. $H_2 + F_2 \longrightarrow$ _____

f. $H_2 + O_2 \longrightarrow$ _____

g. $Al + H_2 \longrightarrow$ _____

h. $H_2 + Br_2 \longrightarrow$ _____

i. $C + H_2 \longrightarrow$ _____

j. $N_2 + H_2 \longrightarrow$ _____

A direct combination reaction appears to be the simplest type of chemical reaction one can imagine. And, in many instances this is true; all that is required is that the two elements be brought into physical contact. Sometimes only a spark is needed to start the reaction. But in other cases temperature and pressure must be carefully controlled to give a good yield, and frequently a catalyst is necessary. Among the reactions in Exercise 2.9, two of these are used in commercial processes, those in which HCl and NH_3 are formed. Hydrogen chloride can be prepared by merely heating the reactants. But the preparation of NH_3 (the Haber process) requires critical temperature and pressure controls as well as the use of a catalyst.

REACTIONS WITH THE HALOGENS

The halogens are found to combine with most metals and many of the nonmetals by direct combination. Normally, they will show a $1-$ oxidation state, but in some instances positive oxidation states are observed. The oxidation states, of course, depend on the relative electronegativities of the bonded atoms. Based on electron configurations, oxidation states of $5+$ and $7+$ would be expected, but $1+$ and $3+$ also commonly occur.

Reactions with Metals.   The halogens do not react with all of the metals by direct combination, but when they do, they form the simple halide salts in which the halogen has an oxidation state of $1-$. The negative oxidation state is, of course, in agreement with the relative electronegativities. In general, the reactions can be represented by

$$2\,M + n\,X_2 \longrightarrow 2\,MX_n$$

where X represents the halogen. Taking a specific example, we see that the results are exactly what we would expect,

$$2\,Ga + 3\,Cl_2 \longrightarrow 2\,GaCl_3$$

If it is assumed that the reactions listed in Exercise 2.11 all proceed by direct combination, the only difficulty you might have in predicting the products of these reactions is that of choosing the proper oxidation state of the metal. So, let's have a quick review.

—————— **EXERCISE** ——————

2.10   Give the expected positive oxidation states for each of the following elements, listing the one you would expect to be most stable (at room temperature) first.

a. Al _____   **b.** Tl _____

c. Ni _____   **d.** Na _____

e. Sn _____   **f.** Cr _____

g. Cd _____   **h.** W _____

i. Bi _____   **j.** Sr _____

Now that the proper oxidation states have been determined, predicting the products of the following reactions should be straightforward.

—————— **EXERCISE** ——————

2.11   Practice makes perfect. Complete the following reactions.

a. $Al + Cl_2 \longrightarrow$ _____   **b.** $Tl + Br_2 \longrightarrow$ _____

c. $Ni + F_2 \longrightarrow$ _____   **d.** $Na + I_2 \longrightarrow$ _____

e. $Sn + Br_2 \longrightarrow$ _____   **f.** $Cr + I_2 \longrightarrow$ _____

g. $Cd + F_2 \longrightarrow$ _____   **h.** $W + Cl_2 \longrightarrow$ _____

i. $Bi + I_2 \longrightarrow$ _____   **j.** $Sr + Br_2 \longrightarrow$ _____

Reactions with Nonmetals.   All of the halogens react with hydrogen to give the expected products, the same products that you saw when we discussed the reactions of hydrogen, namely,

$$H_2 + X_2 \longrightarrow 2\,HX$$

But the reactions with the rest of the nonmetals are not quite so clearcut.

The elements in family **VA** are characterized by the electron configuration ns^2np^3, which leads to the positive oxidation states of $3+$ and $5+$. It is observed that P, As, Sb, and Bi all react by direct combination with the halogens to form the trihalides with the **VA** elements showing the $3+$ oxidation state. In general, we find that

$$2\,M + 3\,X_2 \longrightarrow 2\,MX_3$$

With excess halide, the reaction will continue on to the pentahalide:

$$2\,M + 5\,X_2 \longrightarrow 2\,MX_5$$

Thus if the metal is kept in excess the trihalides are obtained, and if the halogen is kept in excess the pentahalides are obtained. As a specific example, consider

$$2\,As + 3\,Cl_2 \longrightarrow 2\,AsCl_3$$

and

$$2\,As + 5\,Cl_2(excess) \longrightarrow 2\,AsCl_5$$

Actually, in some cases, direct combination is not the best way to prepare the compound, but direct combination reactions do generally work. In the case of the pentahalides, only PBr_5 occurs among the bromides and no pentaiodides occur. It is nice to know these facts, but for our purposes it is more important that you recognize that the halides of the elements in family **VA** will have the formulas MX_3 and MX_5 and that, in general, they can be made by direct combination reactions.

——————   **EXERCISE**   ——————

2.12 As always, the answers are easy if you understand. Complete the following reactions.

a. $H_2 + Br_2 \rightarrow$ _____

b. $Sb + Cl_2(excess) \rightarrow$ _____

c. $P_4 + I_2 \rightarrow$ _____

d. $As + Br_2 \rightarrow$ _____

e. $P_4 + Cl_2(excess) \rightarrow$ _____

f. $Sb + I_2(excess) \rightarrow$ _____

g. $As + Cl_2 \rightarrow$ _____

h. $H_2 + I_2 \rightarrow$ _____

i. $As + F_2(excess) \rightarrow$ _____

j. $Bi + Br_2 \rightarrow$ _____

Members of family **VIA** can react with the halogens by direct combination, but generally direct combination is not the best way to make the halides of this family. If this approach is used, unusual temperatures are often needed. It is also true that many of the halides expected from the electron configurations do not exist, and further, a large number of halides other than those predictable from oxidation states do, in fact, occur. From the family electron configuration, ns^2np^4, we would expect oxidation states of $4+$ and $6+$. And we do find the tetrahalides are generally formed, with the **VIA** elements showing a $4+$ oxidation state and the halogen showing a $1-$ oxidation state. For instance,

$$Se + 2\,Br_2 \longrightarrow SeBr_4$$

Looking at the compounds that you would predict and also do occur, we find

$$
\begin{array}{lll}
SF_4 & SeF_4 & TeF_4 \\
SF_6 & SeF_6 & TeF_6 \\
SCl_4 & SeCl_4 & (TeCl_4)_4 \\
 & SeBr_4 & (TeBr_4)_4 \\
 & & (TeI_4)_4
\end{array}
$$

It is to be noted that hexafluorides of three of the members of family **VIA** (S, Se, and Te) can be made by direct combination, but we observe that the hexahalides ($6+$ oxidation state) occur only with fluorine, the most electronegative element.

Again the question arises as to what should be expected of you as a student. Certainly, it is simple enough to write down the reasonable products for these reactions, but in order to obtain the products that actually occur, you would have to memorize the list of family **VIA** halides. It would be advisable to note the general trends among these halides, but as we have continued to stress, the emphasis in this book will be on learning to predict reasonable products.

_____   **EXERCISE**   _____

2.13   In spite of what was just stated, test your memory by circling the following compounds that might be made by direct combination reactions. Don't worry if you miss a few.

a. $SeBr_6$

b. SCl_4

c. TeF_6

d. SCl_6

e. SI_4

f. TeI_6

g. $SeCl_4$

h. SF_6

i. SeI_4

j. $TeCl_6$

Interhalogen Compounds.   The interhalogens are compounds containing two different halogens such as ICl_3 or ClF. Obviously, one of the halogens must have a positive oxidation state and the other a negative oxidation state. Which will be positive and which will be negative is easily determined from the trend in electronegativity, fluorine being the most electronegative and iodine being the least electronegative among the halogens. Based on the family electron configuration, ns^2np^5, we would predict $7+$ and $5+$ for the positive oxidation states of the halogens. Actually, among the interhalogens, we observe oxidation states of $1+$, $3+$, $5+$, and $7+$ for the more electropositive element in the compound.

All of the interhalogens that exist can be made by direct combination reactions, but a knowledge of which ones actually do occur requires memorization. However, it will be helpful in predicting the products of a direct combination reaction between the halogens to note that (a) they exist, (b) the oxidation states are $1+$, $3+$, $5+$, and $7+$ for the electropositive member, and (c) only the fluorides of the $5+$ and $7+$ compounds occur.

The simplest interhalogens are of the type XX', where X represents one halogen and X' represents the other. All of the interhalogens of this type can be made except IF. Thus, ICl, IBr, BrF, BrCl, and ClF all occur. Here we observe a 1+ oxidation state for the more electropositive element and, of course, a 1− oxidation state for the more electronegative member.

Of the XX_3' type interhalogens, only ClF_3, BrF_3, and ICl_3 occur. Here the more electropositive element has a 3+ oxidation state. For the higher oxidation states, only BrF_5, IF_5, and IF_7 are known. Again we note that, although these compounds can be made by direct combination, it is not always simply a matter of just bringing the elements together. Temperatures, pressures, and concentrations must, in some cases, be carefully controlled.

――――――   **EXERCISE**   ――――――

2.14 Try something a little different this time. List the interhalogen compounds of each of the following types:

a. XX' _____

b. XX_3' _____

c. XX_5' _____

d. XX_7' _____

―――――――――――――

NOBLE GASES AND MISCELLANEOUS EXAMPLES

The noble gases represent a particularly unique family of the elements. From the time of their discovery in the last decade of the nineteenth century until 1962, they were considered to be chemically inert. It was for this reason that they were earlier classified as the inert gases. Even now, relatively few compounds have been prepared with the noble gases and these primarily with xenon. No stable compounds have been made with helium, neon, or argon. The known compounds are only with the most electronegative elements or polyatomic anions, and only fluorine reacts by direct combination. Three fluorides of xenon have been prepared by direct combination with the product depending on the ratio of F_2 to Xe, the pressure, and the temperature. Unfortunately, they cannot be predicted on the basis of electron configurations. They are XeF_2, XeF_4, and XeF_6, and, except for these three cases, the noble gases have not been observed to react by direct combination.

It was pointed out earlier in this chapter that many direct combination reactions do, in fact, occur that would not be predicted by means of oxidation states. And, it was also pointed out that many binary compounds that do exist cannot be made by direct combination. Exercise 2.15 lists a few reactions that do occur, and the products are those one would expect. Particular emphasis has been placed on the nonmetals that were not specifically considered in this chapter.

————   **EXERCISE**   ————

2.15 Complete each of the following reactions.

a. $Sn + S_8 \xrightarrow{\Delta}$ _____   **b.** $Sb + S_8 \xrightarrow{\Delta}$ _____

c. $W + Cl_2 \xrightarrow{\Delta}$ _____   **d.** $Cr + Cl_2 \xrightarrow{\Delta}$ _____

e. $Na + Se \xrightarrow{\Delta}$ _____   **f.** $Bi + S_8 \xrightarrow{\Delta}$ _____

g. $Mo + O_2 \xrightarrow{\Delta}$ _____   **h.** $Si + S_8 \xrightarrow{\Delta}$ _____

i. $K + S_8 \xrightarrow{\Delta}$ _____   **j.** $Ca + Se \xrightarrow{\Delta}$ _____

k. $Ba + S_8 \xrightarrow{\Delta}$ _____   **l.** $C + S_8 \xrightarrow{\Delta}$ _____

m. $Rb + Te \xrightarrow{\Delta}$ _____   **n.** $Li + N_2 \xrightarrow{\Delta}$ _____

REVIEW 1

Here we will review each step of the process for predicting the products of a direct combination reaction from the electron configurations of the elements and their electronegativities. You may not need this review, but it would probably be to your advantage to look at various parts of it to convince yourself that you do, in fact, understand the material. In Chapter 1 you have seen how to determine the reasonable formulas of binary compounds, and in Chapter 2 you learned how to use this information to predict what happens when two elements react with each other. In this review the important steps will be taken up in order. It should be remembered that these procedures can lead only to reasonable products but not necessarily those actually observed. Not all binary compounds can be made by direct combination reactions, and in a few instances quite unexpected products will be formed. Specific examples of unpredictable reactions will, of course, require memorization. This we will avoid whenever possible.

ELECTRON CONFIGURATIONS

Several different means of representing electron configurations have been given, and you should be familiar with all of them. It is extremely important that you use the periodic table as a guide to the configurations rather than any mnemonic devices. In the following exercises we will begin with simple configurations and progress to those that are more involved.

_____ EXERCISES _____

I.1 Use the periodic table to determine the electron configurations for the following.

	Complete	Shorthand	Family	Orbital
				$\downarrow$
a. Na	$1s^2 2s^2 2p^6 3s^1$	$[Ne]\ 3s^1$	ns^1	ns
b. Si				
c. P				
d. Ar				
e. V				
f. As				
g. Ga^{3+}				
h. Sc^{3+}				
i. Se^{2-}				
j. Cl^-				

I.2   For each of the following elements, give all the other elements that you would expect to have similar properties.

a. Na ___Li, K, Rb, Cs, Fr___   **b.** Mo _____

c. Pr _____   **d.** Ge _____

e. I _____   **f.** Zr _____

g. Kr _____   **h.** Tl _____

i. Hg _____   **j.** Te _____

I.3   You were taught a general classification of the elements in Chapter 1. Using this approach, see what you can do with these.

a. Na ___representative___   **b.** Ni _____

c. Rn _____   **d.** Pm _____

e. As _____   **f.** Cm _____

g. W _____   **h.** In _____

i. Ru _____   **j.** Kr _____

OXIDATION STATES

We have presented two different methods for the determination of oxidation states. The first involved an element in a compound. For instance, what is the oxidation state of carbon in C_2H_6? The second method was based on the electron configurations. It is this latter approach that is of concern in predicting the results of direct combination reactions. For the representative elements and the lanthanides, the procedure is quite simple. On the other hand, it is not so simple for the transition elements. You may use Table 1.1 to look up the oxidation states of the transition elements, but you must figure out those of the representative elements and the lanthanides on your own.

———— EXERCISES ————

I.4   Based on the electron configurations, give all of the expected oxidation states for the following elements:

a. Mg _____   b. Br _____

c. Se _____   d. Rb _____

e. Nd _____   f. Tl _____

g. Tm _____   h. Al _____

i. Ra _____   j. In _____

k. Ne _____   l. Sb _____

m. Ba _____   n. Sn _____

o. S _____

I.5   Using your memory, family similarities, and/or Table 1.1, give the common oxidation states for

a. Sc _____   b. Ni _____

c. Fe _____   d. Ag _____

e. Cd _____   f. V _____

g. Zr _____   h. Co _____

i. Mo _____   j. Mn _____

ELECTRONEGATIVITY

Binary compounds can be formed between a metal and a nonmetal and between two nonmetals, but such compounds are not formed between two metals. The combination of two metals generally results in an alloy. These will not be considered in this discussion.

   In a binary compound, one of the elements will be more electronegative than the other. By definition, the more electronegative species tends to attract electrons preferentially to itself, thereby acquiring a negative oxidation state. The other species in the compound will have a positive oxidation state. Thus, in constructing a binary compound, it is necessary to determine which is the more electronegative element and assign it a negative oxidation state. The general trend in electronegativities is shown in Figure 1.5, with fluorine having the highest value. Generally, among the representative elements, the effect is more pronounced in a period (left to right) than in a family. However, if there are conflicting trends, you may find it necessary to refer to the table of electronegativities for the correct answer.

——————   **EXERCISE**   ——————

I.6   For each of the following pairs of elements, indicate which is the more electronegative.

a. Cl or I _____   **b.** S or P _____

c. Mg or Cl _____   **d.** Cr or N _____

e. Na or K _____   **f.** Pb or Si _____

g. Li or I _____   **h.** S or I _____

i. Tl or S _____   **j.** Te or Cl _____

k. As or H _____   **l.** Na or H _____

m. C or H _____   **n.** P or H _____

o. Te or H _____

FORMULAS OF BINARY COMPOUNDS

You now have the necessary information to write down reasonable formulas of binary compounds. Using indium and sulfur as an example, the procedure is as follows:

1. Determine which of the two elements is the more electronegative. Ans. $\underline{S}$
2. Determine the expected oxidation states of each element.
   Ans. $\underline{In: 1+, 3+}$ and $\underline{S: 2-, 4+, 6+}$
3. Assign the more electronegative species a negative oxidation state and the more electropositive species a positive oxidation state(s). If more than one positive oxidation state can occur, then more than one compound will be reasonable.
   Ans. $\underline{In: 1+, 3+}$ and $\underline{S: 2-}$
4. Write down the compound with the more electropositive species first (there are a few exceptions to this rule, such as NH_3 and CH_4). It might be helpful to put the oxidation state above each element. Ans. $\underline{\overset{1+\ 2-}{In\ \ S}}$ and $\underline{\overset{3+\ 2-}{In\ \ S}}$
5. Balance the charges with the appropriate subscripts. Ans. $\underline{In_2S}$ and $\underline{In_2S_3}$

─────── **E X E R C I S E** ───────

I.7   Give the compounds that you would predict to be formed between

a. Tl and O _____   **b.** C and Br _____

c. Na and N _____   **d.** Al and S _____

e. Bi and N _____   **f.** Cr and Cl _____

g. Pm and S _____   **h.** Pb and F _____

i. P and O _____   **j.** Fe and S _____

═══════════════════════════

DIRECT COMBINATION REACTIONS

Now that you can predict a correct formula for a binary compound, the extension to direct combination reactions is straightforward. You merely write down the reacting elements as they occur in nature and then, for the product, write down the formula of a compound formed between these elements. It is important to recognize that the compound you predict will be a reasonable product but not necessarily the correct product. If you must know the actual result, you can always look up the product of a reaction that is of specific interest to you.

_____ **E X E R C I S E** _____

I.8   Complete the following reactions, and show all reasonable products. It is not necessary to balance the equations.

a. $In + Cl_2 \longrightarrow$ _____   **b.** $Sc + O_2 \longrightarrow$ _____

c. $Li + N_2 \longrightarrow$ _____   **d.** $S_8 + O_2 \longrightarrow$ _____

e. $Cr + Br_2 \longrightarrow$ _____   **f.** $Zn + Cl_2 \longrightarrow$ _____

g. $Al + O_2 \longrightarrow$ _____   **h.** $Na + F_2 \longrightarrow$ _____

i. $Pr + Cl_2 \longrightarrow$ _____   **j.** $Ge + O_2 \longrightarrow$ _____

k. $Te + F_2 \longrightarrow$ _____   **l.** $Sn + I_2 \longrightarrow$ _____

m. $Tl + O_2 \longrightarrow$ _____   **n.** $P_4 + Br_2 \longrightarrow$ _____

o. $Sb + Cl_2 \longrightarrow$ _____   **p.** $Fe + S_8 \longrightarrow$ _____

Now consider a few reactions that do, in fact, tend to take place. You have learned the following guidelines:

1. Among the representative elements, when two positive oxidation states are possible, the higher one is more likely with the lighter members of the family. For instance, boron is exclusively $3+$ and aluminum is found almost only as $3+$, whereas thallium tends to form the $1+$ species.
2. a. Na and Ba react with O_2 to form the peroxide, O_2^{2-}.
   b. K, Rb, and Cs react with O_2 to form the superoxide, O_2^{-}.
   c. All the rest of the metals react with O_2 to give the oxide, O^{2-}.
3. Hydrogen reacts with the **IA** and **IIA** metals to form saline hydrides containing the H^- ion.
4. Fluorine tends to take the element with which it reacts to its highest oxidation state.

_____ **E X E R C I S E** _____

I.9   Give the most likely product for each of the following direct combination reactions.

a. $Ca + O_2 \longrightarrow$ _____   **b.** $Rb + O_2 \longrightarrow$ _____

c. $Al + Cl_2 \longrightarrow$ _____   **d.** $Li + O_2 \longrightarrow$ _____

e. $Ge + Cl_2 \longrightarrow$ _____   **f.** $Tl + O_2 \longrightarrow$ _____

g. $Na + O_2 \longrightarrow$ _____   **h.** $Pb + I_2 \longrightarrow$ _____

i. $Ga + F_2 \longrightarrow$ _____   **j.** $Pb + F_2 \longrightarrow$ _____

k. $Cs + O_2 \longrightarrow$ _____   **l.** $Tl + Br_2 \longrightarrow$ _____

m. $K + H_2 \longrightarrow$ _____   **n.** $P_4 + Cl_2 \longrightarrow$ _____

o. $Al + O_2 \longrightarrow$ _____   **p.** $Ba + O_2 \longrightarrow$ _____

If you have mastered the material in this review chapter, you have made a good start in your study of chemical reactions. Keep up the good work, and we think you will find the subsequent chapters to be of ever increasing interest.

3

The Naming of Common Inorganic Compounds

The names of simple inorganic compounds can readily be determined by learning a few rules and referring to the oxidation states and the positions of the elements in the periodic table. The rules of naming (nomenclature) are dependent on how many elements are present in the compound. Compounds containing two elements are called *binary* compounds, and those containing three elements are called *ternary* compounds. We will consider such compounds as well as some containing four elements. The discussion in this chapter will be limited to the very common compounds that have traditionally been a part of first-year chemistry courses. The nomenclature of acids and bases will be considered in Chapter 5.

BINARY COMPOUNDS

Compounds Between Metals and Nonmetals.   There are two common types of binary compounds. The first type is composed of a metal and a nonmetal, whereas the second type is composed of two nonmetals. The rules for naming a compound that contains a metal and a nonmetal are straightforward. The metal is always named first. The nonmetal is then named with its ending changed to *-ide*. Thus NaCl, common table salt, is called sodium chlor*ide*. The simplest of these compounds to name are those in which the metal exhibits only one oxidation state. The more common of these metals, listed by their places in the periodic table, are as follows:

IA	IIA	IB	IIB
Li	Be		
Na	Mg		
K	Ca		Zn
Rb	Sr	Ag	Cd
Cs	Ba		

_____ **EXERCISE** _____

3.1   Practice on these!

            Name

a. NaF   <u>sodium fluoride</u>

b. KI   _____

c. MgO   _____

d. Rb_2S _____

e. $BaBr_2$ _____

f. KH   _____

g. AgCl   _____

Exception! Two exceptions to the simple -_ide_ ending are the diatomic oxide ions, O_2^{2-} and O_2^-. O_2^{2-} is called peroxide and O_2^- is called super-oxide (see p. 31).

Note the differences.

         Barium oxide is BaO.
         Barium peroxide is BaO_2.
         Sodium oxide is Na_2O.
         Sodium peroxide is Na_2O_2.
         Potassium oxide is K_2O.
         Potassium superoxide is KO_2.

—————   **EXERCISES**   —————

3.2   Now let's try some of these!

Name

a. RbO_2 _____

b. CaO_2 _____

c. CsO_2 _____

d. BaO_2 _____

3.3   Give the names of the following compounds. (Be careful.)

a. K_2O _____   **b.** Na_2O _____

c. Na_2O_2 _____   **d.** KO_2 _____

===

Multiple Oxidation States.   Most of the transition metals and the Group IIIA,[1] IVA, and VA metals exhibit multiple oxidation states. The simple nomenclature previously developed is not sufficient in these cases. Iron, for example, possesses both a 2+ and a 3+ oxidation state. Thus, both $FeCl_2$ and $FeCl_3$ can exist. The simple name *iron chloride* does not distinguish between them.

Three systems have been developed to name such compounds. The most common method is the *Stock* system in which a Roman numeral in parentheses is used to indicate the oxidation state of the metal:

$FeCl_2$ is iron(II) chloride.
$FeCl_3$ is iron(III) chloride.
PbO_2 is lead(IV) oxide.
$CuBr_2$ is copper(II) bromide.
SnF_2 is tin(II) fluoride.

A second method, not commonly used in the naming of compounds containing metals, employs a prefix to indicate the number of atoms of a particular element that are present in that compound. Examples are PbO and PbO_2, where one and two oxygen atoms are present in the respective molecules; PbO is called lead monoxide and PbO_2 is called lead dioxide.

---

[1] Al normally shows only a 3+ oxidation state. See p. 29.

The prefixes are

Number of atoms		Number of atoms	
mono-	1	*penta-*	5
di-	2	*hexa-*	6
tri-	3	*hepta-*	7
tetra-	4	*octa-*	8

The prefix *mono-* for one atom is not in general use, but, as usual, we will see that there are exceptions. Some of the few compounds containing metals that still are named using this method are

$TiCl_4$	titanium tetrachloride
OsO_4	osmium tetroxide
MnO_2	manganese dioxide
PbO	lead monoxide

A third and much older nomenclature system makes use of the suffixes *-ous* and *-ic* to denote the lower and the higher oxidation state of the metal, respectively. Thus

$\overset{2+}{Co}Cl_2$ is cobalt*ous* chloride.

$\overset{3+}{Co}Cl_3$ is cobalt*ic* chloride.

It is often stated that this method should be "allowed to die a natural death." However, so long as chemical companies continue to label their products with this system, you need to be aware of it. Imagine if you were to ask a stockroom attendant for iron(III) sulfate and be told that none is in stock while bottles of ferric sulfate were sitting on the shelves! Even the toothpaste manufacturers refer to tin(II) fluoride as stannous fluoride!

One disadvantage of this system is that it uses Latin names for some of the metals that have been known since ancient times—for example, iron, copper, gold, and tin. These Latin names, along with the forms used for the low and high oxidation states, are

Latin name	Low oxidation state	High oxidation state
Fe ferrum	ferrous	ferric
Cu cuprum	cuprous	cupric
Au aurum	aurous	auric[2]
Sn stannum	stannous	stannic

[2] The first name of the infamous villain Goldfinger is Auric. The author, Ian Fleming, knew his nomenclature.

Thus

> $FeCl_2$ is ferrous chloride.
> $FeCl_3$ is ferric chloride.
> $CuBr_2$ is cupric bromide.
> SnF_2 is stannous fluoride.

A second and much more serious disadvantage of this nomenclature system is that it cannot be used for a metal that has more than two oxidation states. Manganese, for example, can have oxidation states of $2+$, $3+$, $4+$, $6+$, and $7+$. Because of such problems, the use of this method has been discouraged by the international nomenclature rules committee.

———— **EXERCISES** ————

3.4 Name these compounds using the Stock system.

a. $AuCl_3$ _____

b. $NiBr_2$ _____

c. CuI _____

d. $FeBr_3$ _____

e. Cr_2S_3 _____

3.5 Try naming these compounds using the "older" method.

a. Fe_2O_3 _____

b. SnO _____

c. $CuBr_2$ _____

d. Hg_2Cl_2 _____

Compounds Between Nonmetals. In binary compounds in which both elements are nonmetals, the recommended system is the one that employs the prefixes mentioned in the preceding section. The more electronegative nonmetal is assigned the -*ide* ending. Thus

> CO_2 is carbon *dioxide*.
> SO_2, an ingredient in the production of acid rain, is sulfur *dioxide*.
> SO_3, another acid rain ingredient, is sulfur *trioxide*.
> N_2O, commonly called laughing gas, is *di*nitrogen ox*ide*.
> NO_2, a common pollutant originating from automobile exhaust, is nitrogen *dioxide*.
> N_2O_5 is *di*nitrogen *pentoxide*.

Again the prefix *mono-* is rarely used; however, CO, a very common compound, is always called carbon monoxide. The international rules committee discourages the use of the Stock system for these nonmetallic binary compounds.

EXERCISES

3.6 Name these binary compounds. Remember that there is only one recommended method.

a. $SiCl_4$ _____   **b.** HCl _____

c. IF_7 _____   **d.** IBr_3 _____

e. N_2O_3 _____   **f.** PCl_5 _____

g. PCl_3 _____

3.7 Now, for some fun and review, name these. Use any system.

a. MnO _____   **b.** Mn_2O_3 _____

c. MnO_2 _____   **d.** Mn_2O_7 _____

TERNARY COMPOUNDS

The ternary compounds encountered in first-year chemistry usually consist of a *monatomic* ion and a *polyatomic* ion. A polyatomic ion consists of more than one type of atom. The only common positive polyatomic ion dealt with in first-year chemistry is the ammonium ion, NH_4^+. Two typical ternary compounds containing the ammonium ion are

NH_4I ammonium iod*ide*
$(NH_4)_2S$ ammonium sulf*ide*

On the other hand, there are many common negative polyatomic ions. Some of the more common ones are listed below according to the position of the central atom (i.e., the non-oxygen atom in these cases) in the periodic table. The nomenclature rules we will discuss apply to the other polyatomic anions as well.

VIB	VIIB	IVA	VA	VIA	VIIA
CrO_4^{2-}	MnO_4^-	CO_3^{2-}	NO_2^-	SO_3^{2-}	ClO^-
$Cr_2O_7^{2-}$			NO_3^-	SO_4^{2-}	ClO_2^-
					ClO_3^-
			PO_4^{3-}		ClO_4^-

The key to the naming of polyatomic anions is the oxidation state of the central atom. Consider the ions CO_3^{2-}, NO_3^-, PO_4^{3-}, SO_4^{2-}, and CrO_4^{2-}. Note that in each case the central atom is in its highest oxidation state.

Ion	Central atom	Group	Oxidation state
CO_3^{2-}	C	IVA	4+
NO_3^-	N	VA	5+
PO_4^{3-}	P	VA	5+
SO_4^{2-}	S	VIA	6+
CrO_4^{2-}	Cr	VIB	6+

Recall that the highest oxidation state of an element is normally the same as its group number. However, we can determine the oxidation state of the central atom by assigning a 2− oxidation state to each oxygen atom and balancing the sum of these with the charge on the polyatomic ion. This procedure will be reviewed in detail in Chapter 4.

When naming these ions in which the central atom is in its highest oxidation state, we write the name of the central atom with the ending changed to -*ate*:

CO_3^{2-}	carbon*ate* ion
NO_3^-	nitr*ate* ion
PO_4^{3-}	phosph*ate* ion[3]
SO_4^{2-}	sulf*ate* ion
CrO_4^{2-}	chrom*ate* ion

Exception! The $Cr_2O_7^{2-}$ ion presents something of an exception to the rule. Each Cr atom possesses a 6+ oxidation state. However, because there are two Cr atoms, it is called the *dichromate* ion. A few typical

---

[3] The correct name is actually orthophosphate, but the prefix *ortho-* is frequently omitted. Use of this prefix will be discussed in Chapter 4.

ternary compounds using these ions are

KNO_3 potassium nitrate
$CaCO_3$ calcium carbonate
$FeSO_4$ iron(II) sulfate
$BaCr_2O_7$ barium dichromate

Usually the central atom will possess either one or the other of two common oxidation states. So now let's consider ions in which the central atom is in its next lowest oxidation state.

Ion	Central atom	Oxidation state
NO_2^-	N	3+
AsO_3^{3-}	As	3+
SO_3^{2-}	S	4+

When naming these ions we change the ending of the name of the central atom to -*ite*:

NO_2^- nitr*ite* ion
AsO_3^{3-} arsen*ite* ion[4]
SO_3^{2-} sulf*ite* ion

Exceptions! Again we find exceptions to the rule. ClO_3^- is called chlor*ate*. Note that chlorine is *not* in its highest oxidation state in this ion. Here Cl is 5+, whereas its highest state is 7+. This exception occurs because four oxychlorine ions exist in which Cl has the oxidation states 7+, 5+, 3+, and 1+. The middle two states are named chlor*ate* (5+) and chlor*ite* (3+). The 7+ state then is identified by *per*chlor*ate* and the 1+ state by *hypo*chlor*ite*. A similar situation arises with the other halogens. MnO_4^- (*per*mangan*ate*) and the much less common MnO_4^{2-} (mangan*ate*) are also exceptions for the same general reasons.

Finally, the diatomic ions CN^- and OH^- have -*ide* endings. CN^- is called cyan*ide*. OH^- is called hydrox*ide*.

---

[4] The PO_3^{3-} ion would be the ideal example to use here, but it does not exist. Thus, we have used the analogous ion AsO_3^{3-}. (See p. 92.)

_____ **EXERCISES** _____

3.8 Now fill in the correct ionic names.

a. PO_4^{3-} _____   **b.** SO_3^{2-} _____

c. NO_2^- _____   **d.** $Cr_2O_7^{2-}$ _____

e. ClO_2^- _____   **f.** SO_4^{2-} _____

g. MnO_4^- _____   **h.** ClO_4^- _____

i. CrO_4^{2-} _____   **j.** NO_3^- _____

k. NH_4^+ _____   **l.** ClO_3^- _____

m. CO_3^{2-} _____   **n.** AsO_3^{3-} _____

3.9 Now reverse the process. Give the formula for each of the following polyatomic ions.

a. nitrate _____   **b.** sulfate _____

c. dichromate _____   **d.** hydroxide _____

e. chlorate _____   **f.** arsenite _____

g. ammonium _____   **h.** chromate _____

i. nitrite _____   **j.** orthophosphate _____

k. perchlorate _____   **l.** permanganate _____

m. sulfite _____   **n.** carbonate _____

3.10   In this exercise, use the Stock system when appropriate.

Name

a. Na_3AsO_3 _____

b. $Ba(OH)_2$ _____

c. $Zn(NO_3)_2$ _____

d. $FeSO_3$ _____

e. $CuCO_3$ _____

f. $Ca_3(PO_4)_2$ _____

g. Ag_2SO_4 _____

h. $Hg(ClO_3)_2$ _____

_____

COMPOUNDS WITH FOUR ELEMENTS

The common ternary ions are derived from $CO_3{}^{2-}$ and $SO_4{}^{2-}$ by the addition of a hydrogen ion and from $PO_4{}^{3-}$ by the addition of one or more hydrogen ions. Thus

$HCO_3{}^-$ is the hydrogen carbonate ion.
$HSO_4{}^-$ is the hydrogen sulfate ion.
$HPO_4{}^{2-}$ is the hydrogen orthophosphate ion.
$H_2PO_4{}^-$ is the dihydrogen orthophosphate ion.

Typical compounds containing four elements are

$NaHCO_3$, which is sodium hydrogen carbonate[5]
KH_2PO_4, which is potassium dihydrogen orthophosphate

Compounds containing four elements may also be derived from two different polyatomic ions. Thus

$(NH_4)_2SO_4$ is ammonium sulfate.
$(NH_4)_3PO_4$ is ammonium orthophosphate.

_____ **EXERCISE** _____

3.11   It's easy now!

Name

a. $(NH_4)_2CO_3$ _____

b. $Ca(HCO_3)_2$ _____

c. NaH_2PO_4 _____

_____

[5] $NaHCO_3$ has traditionally been called sodium bicarbonate. Use of this name is now discouraged by the international rules committee. Its common name is baking soda.

At this point you have the necessary information to name most of the simple inorganic compounds, those that you are likely to find in a first-year chemistry course. But remember that when possible, you should always predict oxidation states. For elements whose oxidation states do not seem to fit the rules, memorize those of only the very common elements. Finally, only as a last resort should you refer to a table of oxidation states.

_____ **EXERCISES** _____

3.12 Now let's see what you've learned. Name the following compounds using the appropriate system.

a. $RbClO_4$ _____

b. $Ca_3(PO_4)_2$ _____

c. Ag_2CrO_4 _____

d. $NaClO$ _____

e. NH_4CN _____

f. $Zn(IO_3)_2$ _____

g. $Al_2(SO_4)_3$ _____

h. Sb_2S_5 _____

i. $As_2(SO_3)_3$ _____

j. TeO_3 _____

3.13 If you did well on Exercise 3.12, go on to the next chapter. If you feel that you need more practice, work on these.

a. $Fe(OH)_3$ _____

b. $CsHCO_3$ _____

c. $Ni_3(PO_4)_2$ _____

d. $HgCl_2$ _____

e. $Mg(MnO_4)_2$ _____

f. PbF_4 _____

g. $SnCr_2O_7$ _____

h. Cl_2O_7 _____

i. $CoHPO_4$ _____

j. $Al(NO_2)_3$ _____

4

Polyatomic Ions, Dissociation of Salts, and Precipitation Reactions

It is not possible to understand and be able to predict the products of chemical reactions if you are not familiar with the polyatomic ions and their behavior. For instance, consider the reaction

$$AgNO_3 + NaCl \xrightarrow{H_2O} products$$

One might wonder why products such as N_2, O_2, or NOCl are not formed, and it is true that in redox (oxidation-reduction) reactions, a variety of products of this type might occur. Such reactions will be discussed in Chapter 8. But in many reactions the polyatomic ion maintains its identity throughout the reaction. In the above case, the products are simply[1]

$$AgNO_3(s) + NaCl(s) \xrightarrow{H_2O} Na^+(aq) + NO_3^-(aq) + AgCl(s)$$

The NO_3^- group is a typical polyatomic anion, and the ability to predict reaction products requires a knowledge of such ions.

POLYATOMIC ANIONS

All but one of the common polyatomic ions are anions, the positive polyatomic ion being the ammonium ion, NH_4^+. We have seen some of the more

---

[1] The symbols (s), (aq), (l), and (g) have their usual meaning of solid, aqueous, liquid, and gas, respectively.

common polyatomic anions and learned their names in Chapter 3. Here we will obtain a better understanding of their behavior.

Oxidation States in Polyatomic Ions.   In Chapter 1 we learned how to determine the oxidation state of an element in a compound. We also learned that the actual oxidation state may or may not be the same as would have been predicted from the electron configuration. For instance, in H_2O_2, it was concluded that oxygen has an oxidation state of $1-$, rather than the $2-$ that one would predict from its electron configuration. However, in all of the polyatomic ions given in Chapter 3 (except for ClO^- and ClO_2^-), the oxidation states do turn out to be those one would predict.

   To obtain the oxidation state of the central element in a polyatomic ion, we assign the normal oxidation states to the common elements such as oxygen, chlorine, hydrogen, etc., and then choose the appropriate oxidation number for the central atom that will result in the observed ionic charge. For example, consider the sulfate ion SO_4^{2-}. We can see that the sulfur has an oxidation state of $6+$ from the following calculation:

$$
\begin{array}{lr}
\text{oxygen:} & -2 \times 4 = -8 \\
\text{sulfur:} & ? \times 1 = \underline{+6} \\
\text{resultant charge:} & -2
\end{array}
$$

   As another example, consider the dichromate ion, $Cr_2O_7^{2-}$. The oxidation state of each Cr atom can be seen to be $6+$:

$$
\begin{array}{lr}
\text{oxygen:} & -2 \times 7 = -14 \\
\text{chromium:} & +6 \times 2 = \underline{+12} \\
\text{resultant charge:} & -2
\end{array}
$$

   Now let's consider $S_2O_3^{2-}$. The oxidation state of sulfur in this ion does not turn out to be one of those normally expected. However, it is readily determined as follows:

$$
\begin{array}{lr}
\text{oxygen:} & -2 \times 3 = -6 \\
\text{sulfur:} & ? \times 2 = \underline{x} \\
\text{resultant charge:} & = -2
\end{array}
$$

Obviously x must be $+4$, and each sulfur, therefore, has an oxidation state of $2+$.

EXERCISES

4.1 Now let's see if you have the idea. Determine the oxidation state of the underlined element.

a. $\underline{S}O_3^{2-}$ _____

b. $\underline{N}H_4^{+}$ _____

c. $\underline{P}O_4^{3-}$ _____

d. $\underline{N}O_2^{-}$ _____

e. $\underline{N}O_3^{-}$ _____

f. $\underline{S}O_4^{2-}$ _____

g. $\underline{Mn}O_4^{-}$ _____

h. $\underline{P}O_3^{-}$ _____

i. $\underline{S}_2O_7^{2-}$ _____

j. $\underline{C}_2O_4^{2-}$ _____

k. $\underline{C}_2H_3O_2^{-}$ _____

l. $\underline{S}_2O_3^{2-}$ _____

4.2 If you think you need the extra practice, do some more of the same. Otherwise, read on.

a. $\underline{Cr}O_4^{2-}$ _____

b. $\underline{Cl}O_3^{-}$ _____

c. $\underline{Te}O_4^{2-}$ _____

d. $\underline{Mn}O_4^{-}$ _____

e. $\underline{C}O_3^{2-}$ _____

f. $\underline{Cl}O_2^{-}$ _____

g. $\underline{Al}F_6^{3-}$ _____

h. $\underline{Ag}Cl_2^{-}$ _____

i. $\underline{Cl}O_4^{-}$ _____

j. $\underline{Cr}_2O_7^{2-}$ _____

k. $\underline{Mn}O_4^{2-}$ _____

l. $\underline{Cl}O^{-}$ _____

Periodic Similarities. If a particular polyatomic anion exists for sulfur, it should not be surprising to find that similar ions exist for selenium and tellurium as well. Such parallelism within a family will not always be the case, but using it is a logical approach for predicting which polyatomic ions exist among the less familiar elements. Thus, since the sulfite anion, SO_3^{2-}, exists, we should expect that the selenite anion, SeO_3^{2-}, as well as the tellurite anion, TeO_3^{2-}, also exists.

EXERCISE

4.3 To again correlate nomenclature and periodicity, give the formulas for the following polyatomic ions. If, at first, you can't answer one of these, remember what you know about family similarities. In part a, for example, you should know that the phosphate ion is PO_4^{3-}; thus the arsenate ion probably is AsO_4^{3-}.

a. arsenate _____   **b.** perbromate _____

c. tellurite _____   **d.** silicate _____

e. molybdate _____   **f.** iodate _____

g. stannate _____   **h.** arsenite _____

i. selenate _____   **j.** phosphonium _____

Ortho, Meso, and Meta Forms. It is often found that oxyacids[2] can show various degrees of hydration, and this variability can result in different forms of their polyatomic anions. For instance, two forms of arsenous acid occur, H_3AsO_3 and $HAsO_2$. $HAsO_2$ is called *meta*arsenous acid, and H_3AsO_3 is called *ortho*arsenous acid. The corresponding anions are

Acid	Anion	Name
$HAsO_2$	AsO_2^-	*meta*arsenite ion
H_3AsO_3	AsO_3^{3-}	*ortho*arsenite ion

Note that the oxidation state of arsenic is $3+$ in both forms. It is the $3+$ oxidation state of the arsenic that makes the acid arsenous acid and the corresponding anion the arsen*ite* anion. The two different forms are then distinguished with the prefixes *meta-* and *ortho-*.

Now note that the difference between $HAsO_2$ and H_3AsO_3 is one H_2O molecule. That is, if you add H_2O to $HAsO_2$, you get H_3AsO_3. The difference between the meta and ortho forms is the degree of hydration. $HAsO_2$ shows the smallest possible number of hydrogen and oxygen atoms, and H_3AsO_3 has an equal number of hydrogen and oxygen atoms. The latter represents the maximum extent of hydration. In fact, for a true ortho acid, the formula can be written as $M(OH)_x$. In the case of H_3AsO_3, we can write $As(OH)_3$.

Sometimes, because of historical usage or, perhaps, sloppy terminology, the names of oxyacids deviate from these rules. For phosphor*ic* acid one might expect the following:

[2] Oxyacids will be discussed in more detail in Chapter 5.

Acid	Anion	Prefix
HPO_3	PO_3^-	meta
H_3PO_4	PO_4^{3-}	meso
H_5PO_5	PO_5^{5-}	ortho

However, H_5PO_5 does not exist. As a consequence, H_3PO_4 is referred to as *ortho*phosphor*ic* acid, and the corresponding anion is called the *ortho*phosph*ate* ion. Generally, such a situation does not arise. It is also frequently true that only one form of an oxyacid exists and, consequently, no prefix is used. This is the case for the oxyacids of chlorine, sulfur, and nitrogen. HNO_3, for example, is referred to simply as nitric acid rather than metanitric acid, and the corresponding polyatomic anion is simply the nit*rate* ion.

EXERCISES

4.4 Quite a bit of work here! Given the central atom and its oxidation state, write all possible forms (whether they exist or not) of the oxyacid and its corresponding polyatomic anion, and give the name of each anion.

a. Te^{6+} H_2TeO_4 TeO_4^{2-} metatellurate ion

         H_4TeO_5 TeO_5^{4-} mesotellurate ion

         H_6TeO_6 TeO_6^{6-} orthotellurate ion

b. As^{3+}

c. Sb^{5+}

d. Se^{4+}

e. Al^{3+}

f. Sn^{4+}

4.5 Give the formulas of all possible forms of the oxyacid and the corresponding polyatomic anion for I^{5+}.

4.6 For a review, give the complete name of each of the following polyatomic anions.

a. BO_3^{3-} _____

b. PO_3^- _____

c. BrO_3^- _____

d. SeO_4^{4-} _____

e. AlO_2^- _____

f. AlO_3^{3-} _____

g. GeO_4^{4-} _____

h. AsO_3^- _____

SALTS

A *salt* can be defined as an ionic compound containing a cation other than $H^+(H_3O^+)$[3] and an anion other than O^{2-}, O_2^{2-}, O_2^-, or OH^-. From this definition we can readily see that salts can contain polyatomic ions as well as simple ions. Some typical examples of salts are

$NaCl$	sodium chloride
$Ca_3(PO_4)_2$	calcium orthophosphate
KNO_3	potassium nitrate
$NaNO_2$	sodium nitrite
$BaSO_4$	barium sulfate
Rb_2CO_3	rubidium carbonate
$(NH_4)_2SO_3$	ammonium sulfite
$Al_2(CrO_4)_3$	aluminum chromate
NH_4ClO_3	ammonium chlorate

[3] Actually, a number of other cations must also be excluded, such as I^+ and NO_2^+, but these are not commonly encountered in first-year chemistry.

The question of whether a compound is or is not ionic can be complicated. A correct definition of an ionic compound should be based on its properties, but this is of little help in situations such as this. You don't have time to check on the properties of each compound that is presented. However, there are several rules of thumb that can be used to help us make reasonable guesses. One of these is the Pauling criterion that a difference in electronegativity of two or more between the bonded atoms will result in an ionic bond. Using this approach, we find from the electronegativity table in the back of the book that $AlCl_3$ is covalent whereas NaCl is ionic. Thus NaCl is a salt whereas $AlCl_3$ is not.

	EN		EN
Al:	1.5	Na:	0.9
Cl:	3.0	Cl:	3.0
$\Delta EN = 1.5$		$\Delta EN = 2.1$	

Another general rule is that compounds formed between elements in families IA and IIA and those in families VIA and VIIA will be ionic. A third rule is that compounds containing polyatomic anions and a cation other than H^+ are ionic.

One should keep in mind, however, that rules of thumb are only guidelines, and many exceptions will occur.

—————— **EXERCISES** ——————

4.7 Which of the following compounds are salts? Use the table of electronegativities when necessary, and circle the correct answers.

a. H_2CO_3 **b.** $CaCl_2$

c. C_6H_6 **d.** N_2O_5

e. Na_2O_2 **f.** C_2H_5OH

g. ICl_5 **h.** $Al(NO_3)_3$

i. $ZnSO_4$ **j.** IBr

k. Ga_2S_3 **l.** $TlNO_3$

4.8 Again, if you feel you need more practice identifying salts, circle the salts in the list below.

a. PCl_3 **b.** $SrTeO_3$

c. NH_4OH **d.** $Bi(ClO_4)_3$

e. HCl **f.** $HgCl_2$

g. $SeCl_4$ **h.** $Ca(OH)_2$

i. SeO_2 **j.** $Al_2(SO_4)_3$

k. TeI_4 **l.** $Ba(OH)_2$

Dissociation of Salts. When a salt dissolves, it is assumed to ionize completely. Thus, we have

$$NaCl(s) \xrightarrow{H_2O} Na^+(aq) + Cl^-(aq)$$

Here we see that two ions are formed in solution for every one unit of NaCl put into the solution. For $CaCl_2$, we obtain three ions, as can be seen here:

$$CaCl_2(s) \xrightarrow{H_2O} Ca^{2+}(aq) + 2\ Cl^-(aq)$$

Now if we choose a compound containing a polyatomic anion, such as $Ca(NO_3)_2$, we find that the anion maintains its identity:

$$Ca(NO_3)_2(s) \xrightarrow{H_2O} Ca^{2+}(aq) + 2\ NO_3^-(aq)$$

Consequently, in this case we again obtain three ions.
    For a slightly more complex example, consider

$$Na_3PO_4(s) \xrightarrow{H_2O} 3\ Na^+(aq) + PO_4^{3-}(aq)$$

Here we have four ions formed in solution. Again note that the polyatomic anion maintains its identity.

--- **EXERCISE** ---

4.9 Give the products of ionization for each of the following salts. Be sure to indicate the numbers of ions formed. Assume that each salt is dissolved in water.

a. $K_2S(s) \xrightarrow{H_2O} \underline{2\ K^+(aq) + S^{2-}(aq)}$

b. $KClO_3(s) \longrightarrow$ _____

c. $BaI_2(s) \longrightarrow$ _____

d. $CsBr(s) \longrightarrow$ _____

e. $K_2Cr_2O_7(s) \longrightarrow$ _____

f. $(NH_4)_2SO_4(s) \longrightarrow$ _____

g. $Rb_2SO_3(s) \longrightarrow$ _____

h. $Al_2(SO_4)_3(s) \longrightarrow$ _____

i. $Ga_2(CO_3)_3(s) \longrightarrow$ _____

j. $Cs_3PO_4(s) \longrightarrow$ _____

k. $KMnO_4(s) \longrightarrow$ _____

l. $RbNO_3(s) \longrightarrow$ _____

m. $(NH_4)_3PO_4(s) \longrightarrow$ _____

n. $CaI_2(s) \longrightarrow$ _____

o. $NH_4Cl(s) \longrightarrow$ _____

p. $Bi(NO_3)_3(s) \longrightarrow$ _____

Precipitation Reactions. Some salts are very soluble in water, whereas many others are only slightly soluble. In practice, we often consider the latter to be insoluble. Solubilities, in fact, can be used to predict the results of some metathetical reactions. In a *metathetical* reaction (metathesis), the cations and anions of different compounds merely trade partners. The precipitation reactions considered here are a type of metathetical reaction. An example of such a reaction is

$$AgNO_3(s) + NaCl(s) \xrightarrow{H_2O} NaNO_3(aq) + AgCl(s)$$

Since salts ionize in solution, one might expect nothing more than

$$AgNO_3(s) + NaCl(s) \xrightarrow{H_2O} Ag^+(aq) + NO_3^-(aq) + Na^+(aq) + Cl^-(aq)$$

If this were the case, then no reaction would have actually taken place. However, there are circumstances under which a reaction does, in fact, occur. One of these circumstances involves the formation of an insoluble compound. In the example we are considering here, the reaction proceeds because AgCl is relatively insoluble and precipitates from the solution, leaving Na^+ and NO_3^- ions. These can be obtained in the form of $NaNO_3(s)$ by removing the solid $AgCl$ and then evaporating the water. The pertinent equations are

$$AgNO_3(s) + NaCl(s) \xrightarrow{\;H_2O\;} AgCl(s) + Na^+(aq) + NO_3^-(aq)$$

$$Na^+(aq) + NO_3^-(aq) \xrightarrow{\text{evaporation}} NaNO_3(s) + H_2O(g)$$

In this particular reaction, it is important that $AgNO_3$ by itself is soluble in water and $NaCl$ by itself is soluble in water, but that $AgCl$ is insoluble whereas $NaNO_3$ is soluble. If any one of these conditions were not met, then the reaction would not have taken place. In this respect consider

$$Ba(NO_3)_2(s) + 2\,NaBr(s) \xrightarrow{\;H_2O\;}$$
$$Ba^{2+}(aq) + 2\,NO_3^-(aq) + 2\,Na^+(aq) + 2\,Br^-(aq)$$

Here $Ba(NO_3)_2$, $BaBr_2$, $NaNO_3$, and $NaBr$ are all soluble in water. Consequently, only the four ion types shown are obtained, and no precipitation reaction is observed.

Obviously, in order to predict the results of reactions of this type, the solubilities of the salts must be known. Following is a list of solubilities of the more common salts. It is helpful to commit this table to memory.

Solubilities of Common Salts in Water

1. Salts of group **IA** elements and NH_4^+ are soluble.
2. Nitrates, chlorates, perchlorates, and acetates are soluble.
3. Chlorides, bromides, and iodides are soluble except for those of Pb^{2+}, Ag^+, and Hg_2^{2+}.
4. Sulfates are soluble except for those of Ba^{2+}, Sr^{2+}, Pb^{2+}, Hg_2^{2+}, Ag^+, and Ca^{2+} ($CaSO_4$ and Ag_2SO_4 are slightly soluble).
5. Sulfides are insoluble except for those of elements in groups **IA**, and **IIA**, and of NH_4^+.
6. Phosphates, carbonates, and chromates are insoluble except for those of elements in group **IA** and those of NH_4^+.

EXERCISE

4.10 Which of the following compounds are considered to be insoluble in water? Use the solubility table if necessary.

a. K_2S **b.** $AgBr$

c. NH_4NO_3 **d.** $Al_2(CO_3)_3$

e. $BaSO_4$ **f.** $SrCl_2$

g. $CrPO_4$ **h.** $Pb(ClO_4)_2$

i. $HgSO_4$ **j.** $PbCl_2$

k. $ZnSO_4$ **l.** Cs_2CO_3

m. Na_3PO_4 **n.** Ag_2S

o. Hg_2I_2

If solubility is to be the driving force for a precipitation reaction between two salts, then it is necessary that one of the products be insoluble. Consider the reaction

$$BaCl_2(s) + Na_2SO_4(s) \xrightarrow{H_2O}$$

First we note that both $BaCl_2$ and Na_2SO_4 are soluble in water. Thus it is possible to make up a solution of each compound. Now if the two solutions are mixed, the four ion types

$$Ba^{2+}, \quad Cl^-, \quad Na^+, \quad SO_4^{2-}$$

will all come in contact with one another. But since $BaSO_4$ is insoluble, it will precipitate from the solution, leaving Na^+ and Cl^- ions. The $BaSO_4$ can be removed by filtering the solution, and the NaCl can then be recovered by evaporation of the remaining solution. The reaction is

$$BaCl_2(s) + Na_2SO_4(s) \xrightarrow{H_2O} 2\,NaCl(aq) + BaSO_4(s)$$

or

$$BaCl_2(s) + Na_2SO_4(s) \xrightarrow{H_2O} 2\,Na^+(aq) + 2\,Cl^-(aq) + BaSO_4(s)$$

_____ **EXERCISES** _____

4.11 Now let's see how much you have learned. Give the results of each of the following reactions. If no precipitate occurs, write NR (for no reaction). Assume all of the reactants are dissolved in water.

a. $Hg(NO_3)_2 + Na_2S \longrightarrow$ $\underline{2\,Na^+(aq) + 2\,NO_3^-(aq) + HgS(s)}$

b. $Na_2CO_3 + SnCl_2 \longrightarrow$ _____

c. $2\,AgNO_3 + BaCl_2 \longrightarrow$ _____

d. $Ag_2SO_4 + Ba(NO_3)_2 \longrightarrow$ _____

e. $3\ AgNO_3 + Na_3PO_4 \longrightarrow$ _____

f. $(NH_4)_2CO_3 + 2\ RbCl \longrightarrow$ _____

g. $(NH_4)_2CO_3 + SrCl_2 \longrightarrow$ _____

h. $3\ Sn(ClO_3)_2 + 2\ Na_3PO_4 \longrightarrow$ _____

4.12 If you need more practice, give the results of these reactions.

a. $Na_2S + 2\ AgClO_4 \longrightarrow$ _____

b. $Na_2S + 2\ RbNO_3 \longrightarrow$ _____

c. $ZnCl_2 + K_2SO_4 \longrightarrow$ _____

d. $Hg(NO_3)_2 + 2\ CsCl \longrightarrow$ _____

e. $Rb_2S + Fe(ClO_3)_2 \longrightarrow$ _____

f. $(NH_4)_2CrO_4 + Pb(NO_3)_2 \longrightarrow$ _____

g. $3\ Na_2SO_4 + 2\ Al(NO_3)_3 \longrightarrow$ _____

h. $CdCl_2 + Sn(NO_3)_2 \longrightarrow$ _____

You have now completed chapters on nomenclature and on the behavior of ionic compounds. Your next task then is to complete another review section, one in which you will be expected to predict the results of ionic reactions as well as name both the reactants and the products of these reactions.

REVIEW II

In Chapter 3 you learned how to name common inorganic compounds. In Chapter 4 you learned how to predict the products of precipitation reactions. Of course, before you could actually succeed in these predictions, you first had to familiarize yourself with the nature of polyatomic ions and the dissociation of salts. In this review chapter, all of these important concepts will again be presented in the order in which you learned them. If you feel that you understand one of the topics, go on to the next one. If you feel confident about all of this material, move on to Chapter 5.

NOMENCLATURE

Binary Compounds. We have presented several systems of nomenclature for binary compounds. You should always use the particular system recommended by the international rules committee. You need, however, to be aware of the system that uses the -*ous* and -*ic* endings denoting the lowest and highest oxidation states, respectively, so that you can read the labels on chemical bottles.

When naming compounds, avoid giving unnecessary information. Don't, for example, call $ZnCl_2$ zinc(II)chloride, since zinc has only one oxidation state. Roman numerals, remember, are used only for metals that exhibit multiple oxidation states.

EXERCISES

II.1 Now, let's see what you've remembered. Name the following binary compounds. Use the recommended name in each case.

a. $ZnCl_2$ _____

b. SrO _____

c. FeI_2 _____

d. Cu_2S _____

e. PbF_4 _____

f. N_2O _____

g. PCl_3 _____   **h.** SF_6 _____

i. N_2S_5 _____   **j.** S_2Cl_2 _____

II.2   You should not forget the exceptions. In this exercise the normal oxide is given in the left column, whereas the corresponding peroxide or superoxide is given in the right column. Name these compounds.

a. Na_2O _____ Na_2O_2 _____

b. K_2O _____ KO_2 _____

c. Cs_2O _____ CsO_2 _____

d. BaO _____ BaO_2 _____

_____

Ternary Compounds.   Here you must remember that the common ternary compounds encountered in general chemistry contain a monatomic ion and a polyatomic ion. All you have to do then to name the compound is name the ions. To make it even easier, the only common polyatomic cation is the ammonium ion (NH_4^+). The names of the polyatomic anions are based on the oxidation state of the central atom.

_____ **EXERCISES** _____

II.3   Name the following ternary compounds. Refer to the periodic table if necessary.

a. NH_4I _____   **b.** $NaNO_3$ _____

c. $CuSO_4$ _____   **d.** $PbCO_3$ _____

e. KNO_2 _____   **f.** $Cd_3(PO_4)_2$ _____

g. $Hg(ClO_3)_2$ _____   **h.** $FeSO_3$ _____

II.4   Again, you should remember the exceptions. Name these ions.

a. $Cr_2O_7^{2-}$ _____   **b.** ClO^- _____

c. ClO_4^- _____   **d.** CN^- _____

e. OH^- _____   **f.** ClO_2^- _____

_____

Compounds with Four Elements.   The common inorganic compounds containing four elements are either formed from two polyatomic ions or derived from ternary ions produced by adding a hydrogen ion to CO_3^{2-} or SO_4^{2-} or by adding one or two hydrogen ions to PO_4^{3-}. For compounds formed from two polyatomic ions, just name the ions. The ternary ions with one hydrogen ion are called hydrogen carbonate, hydrogen sulfate, and hydrogen orthophosphate ions, respectively. The $H_2PO_4^-$ ion is called the dihydrogen orthophosphate ion.

_____ **EXERCISES** _____

II.5   Name the following compounds containing two polyatomic ions.

a. NH_4NO_3 _____

b. $(NH_4)_3PO_4$ _____

c. $(NH_4)_2SO_4$ _____

II.6   Name these ions.

a. SO_4^{2-} _____

b. HSO_4^- _____

c. PO_4^{3-} _____

d. HPO_4^{2-} _____

e. CO_3^{2-} _____

f. HCO_3^- _____

g. $H_2PO_4^-$ _____

h. $H_2PO_3^-$ _____

II.7   Now, name these compounds.

a. K_3PO_4 _____

b. K_2HPO_4 _____

c. KH_2PO_4 _____

II.8   Don't forget these! Give the oxidation state of the underlined element.

a. $\underline{S}O_4^{2-}$ _____

b. $\underline{N}O_3^-$ _____

c. $\underline{Mn}O_4^-$ _____

d. $\underline{Cl}O_2^-$ _____

e. $\underline{S}O_3^{2-}$ _____

f. $H_2\underline{P}O_4^-$ _____

_____

Periodic Similarities.   The fact that periodic similarities exist certainly makes it easier to determine the formulas of less common polyatomic ions.

_____ EXERCISE _____

II.9 Determine the formulas and the names of the following polyatomic ions. Use the periodic table for this one. Assume the oxidation state of each central atom is $4+$.

	Central atom	Formula of ion	Name of ion
a.	C	$CO_3{}^{2-}$	_____
b.	Si	_____	_____
c.	Ge	_____	_____
d.	Sn	_____	stannate

Ortho-, Meso-, and Meta- Forms. Oxyacids often show various degrees of hydration and thus may exhibit different forms. Consider the oxyacid H_2TeO_4 and its corresponding anion, $TeO_4{}^{2-}$. This acid has two other possible forms. The addition of one water molecule of hydration produces the acid H_4TeO_5 and its corresponding anion, $TeO_5{}^{4-}$. The addition of a second water molecule produces the acid H_6TeO_6. H_2TeO_4 represents the least degree of hydration, H_4TeO_5 represents the intermediate degree of hydration, and H_6TeO_6 represents the maximum degree of hydration. Thus, H_2TeO_4, H_4TeO_5, and H_6TeO_6 are the *meta-*, *meso-*, and *ortho-* forms, respectively, of telluric acid. It should be noted that in many cases, such as for arsenous acid, only two forms can exist, the *meta-* and *ortho-*. Of course, when only one form of the acid exists, as for HNO_3, there is no need for a prefix. Remember that the maximum degree of hydration allowed is for the true *ortho-* form, whose formula can be written as $M(OH)_x$.

_____ EXERCISE _____

II.10 Fill in the blanks. As always, refer to the periodic table.

	Acid	Anion	Prefix	Name of acid
a.	H_2SnO_3	_____	*meta-*	_____
b.	H_4SnO_4	_____	_____	orthostannic
c.	$HAlO_2$	_____	_____	_____
d.	H_3AlO_3	_____	_____	_____
e.	HIO_3	_____	_____	_____

f. H_3IO_4 ____   ____   _____

g. H_5IO_5 ____   ____   _____

It is to be emphasized that the preceding are only possible forms. It is beyond the scope of this book to discuss which of these oxyacids actually do exist.

SALTS

Because of your familiarity with the kinds of ions, both simple and polyatomic, you can now investigate the behavior of compounds composed of these ions. These ionic compounds, with a few exceptions, are called salts.

Things, however, are not as simple as we may like. It can be rather difficult, it turns out, to determine whether or not a given compound is ionic and therefore fits the definition of a salt. You might assume, for example, that PbI_2 is ionic, since it would appear to be composed of lead(II) ions and iodide ions. But this is not the case. PbI_2 is a covalent compound. In order to avoid these difficulties, rules of thumb have been developed so that the student may make reasonable guesses as to which compounds are indeed ionic. Remember, however, that whenever rules of thumb are used, there will always be exceptions.

_____ **EXERCISE** _____

II.11   Using the three rules of thumb that you learned in Chapter 4, circle the salts.

a. NaCl   **b.** PbI_2

c. $ZnSO_4$ **d.** MgF_2

e. Rb_2S **f.** $Fe(NO_3)_3$

g. BaO_2 **h.** NaOH

i. Al_2S_3 **j.** Li_3PO_4

k. NH_4NO_2 **l.** Na_2O_2

Dissociation of Salts.   We assume that when salts dissolve in water, they ionize completely. The polyatomic ions, however, retain their identities. Thus we have

$$KBr(s) \xrightarrow{H_2O} K^+(aq) + Br^-(aq)$$

$$NaNO_3(s) \xrightarrow{H_2O} Na^+(aq) + NO_3^-(aq)$$

$$(NH_4)_2SO_4(s) \xrightarrow{H_2O} 2\ NH_4^+(aq) + SO_4^{2-}(aq)$$

———— E X E R C I S E ————

II.12   Give the products of the ionization of each of the following:

a. $NaCl(s) \xrightarrow{H_2O}$ _____

b. $Mg(NO_3)_2(s) \xrightarrow{H_2O}$ _____

c. $FeCl_3(s) \xrightarrow{H_2O}$ _____

d. $(NH_4)_2S(s) \xrightarrow{H_2O}$ _____

e. $K_3PO_4(s) \xrightarrow{H_2O}$ _____

f. $(NH_4)_2SO_4(s) \xrightarrow{H_2O}$ _____

g. $KH_2PO_4(s) \xrightarrow{H_2O}$ _____

Precipitation Reactions.   You may have wondered why, in a book dealing with reaction chemistry, so much attention is given to polyatomic ions, salts, the dissociation of salts, and solubilities. Since the *driving force* of a metathetical reaction is often the production of a precipitate, you need to know the solubility rules for salts. And once you are familiar with solubilities, dissociation behavior, and the stabilities of polyatomic ions during reactions, you will be able to successfully predict the products of precipitation reactions.

Consider the reaction of $BaCl_2$ and Na_2SO_4 in water. Both compounds are soluble in water; thus,

$$BaCl_2(s) + Na_2SO_4(s) \xrightarrow{H_2O} Ba^{2+}(aq) + 2\ Cl^-(aq) + 2\ Na^+(aq) + SO_4^{2-}(aq)$$

But NaCl is soluble, whereas $BaSO_4$ is insoluble; thus,

$$BaCl_2(s) + Na_2SO_4(s) \xrightarrow{H_2O} BaSO_4(s) + 2\ Na^+(aq) + 2\ Cl^-(aq)$$

Barium chloride and sodium sulfate are the reactants. The products are aqueous sodium chloride and a precipitate, which is barium sulfate.

_____ **EXERCISE** _____

II.13 Now predict the products of the following reactions, assuming that all of the reactants are dissolved in water. Also, for a review of nomenclature, give the names of the reactants and the products. Remember to indicate which salt precipitates and which salt remains in solution.

a. $Na_2SO_4 + BaCl_2 \xrightarrow{\text{H}_2\text{O}} BaSO_4(s) + 2\ Na^+(aq) + 2\ Cl^-(aq)$

sodium sulfate + barium chloride → barium sulfate + sodium chloride

b. $ZnCl_2 + K_2SO_4 \longrightarrow$ _____

_____ + _____ → _____ + _____

c. $AgNO_3 + Na_3PO_4 \longrightarrow$ _____

_____ + _____ → _____ + _____

d. $(NH_4)_2SO_4 + Pb(NO_3)_2 \longrightarrow$ _____

_____ + _____ → _____ + _____

e. $Hg_2(NO_3)_2 + NaCl \longrightarrow$ _____

_____ + _____ → _____ + _____

f. $CaS + FeCl_3 \longrightarrow$ _____

_____ + _____ → _____ + _____

g. $Na_2CrO_4 + AgNO_3 \longrightarrow$ _____

_____ + _____ → _____ + _____

h. $(NH_4)_3PO_4 + CuCl_2 \longrightarrow$ _____

_____ + _____ → _____ + _____

If you have mastered this material, you are now ready to begin a study of acid-base reactions. Acid-base chemistry is somewhat challenging, but you are ready to meet that challenge.

5

Acids and Bases

Historically, acids were set apart from other compounds because of their unique properties such as a sour taste and the ability to change the colors of certain vegetable dyes, some of which are still used today as acid-base indicators. There are many compounds which can be classified as acids that occur naturally. Citric acid and acetic acid (vinegar) are among the most common. But in addition to the ability to identify an acid, chemists also sought an insight into the reasons for their unique behavior. This has led to several useful definitions of an acid, the one most commonly used being the *Brønsted-Lowry* definition. This definition is also frequently referred to as the *protonic* definition. Other definitions of acids exist, and one of them, the Lewis definition, will be considered in Chapter 7. The various definitions become particularly important when we recognize that all chemical reactions, according to some views, can be classified as either *acid-base* or *oxidation-reduction*. Thus, if a reaction produces no change in oxidation state, it can be classified as an acid-base reaction according to at least one of the acid-base definitions.

BRØNSTED-LOWRY ACIDS

According to the Brønsted-Lowry definition, an acid is any substance capable of donating a proton. An acid then is a *proton donor*, hence the commonly used name *protonic acid*. It should be pointed out that a substance is not a protonic acid simply because it contains hydrogen. It must be capable of donating one or more hydrogen ions (protons) in a chemical reaction in order to be so classified. For instance, hydrogen chloride (HCl) readily donates a proton in most reactions and in those cases is an acid, whereas methane (CH_4), which contains four hydrogen atoms per molecule, does not give up a proton and therefore is not an acid. As a further example, acetic acid, the acidic component in vinegar, has the formula CH_3COOH. This compound contains four hydrogen atoms, but only one is acidic. This is the conventional way of writing an organic acid, and it is the hydrogen atom bonded to an oxygen atom that is acidic.

In most instances the formula for an inorganic Brønsted-Lowry acid is written with the acidic hydrogen first, as in HCl, H_2SO_4, HNO_3, and H_3PO_4. This is not always the case, but if the formula of a compound starts with H, there is a good chance it is a protonic acid.

The Naming of Brønsted-Lowry Acids. We can begin our discussion of acid nomenclature by separating the Brønsted-Lowry acids into two general classes, *hydroacids* and *oxyacids*. The hydroacids usually contain only two elements, one of which is hydrogen. A notable exception is HCN. Typical examples of hydroacids are HF, HCl, HBr, and HI. More generally, they can be represented as H_nX, where hydrogen atoms are covalently bonded to certain nonmetal atoms. Thus, in addition to the rather commonplace hydroacids just listed, this class also includes such compounds as H_2S and H_2Se.

The hydroacids are very simple to name, and, in fact, you probably already know many of them. The non-hydrogen element is given a *hydro*-prefix and an *-ic* ending, and this is followed by the word acid. Thus, we have

HCl	*hydro*chlor*ic* acid
HI	*hydr*iod*ic* acid[1]
H_2S	*hydro*sulfur*ic* acid
HCN	*hydro*cyan*ic* acid

The oxyacids are much more varied than the hydroacids, some common examples being HNO_3, H_2SO_4, $HClO_3$, and H_3PO_4. With oxyacids we are dealing with covalent compounds in which the acidic hydrogen atoms are attached to oxygen atoms that are, in turn, attached to the central atom. The central atom is most commonly, but not necessarily, a nonmetal.

---

[1] HI is an exception in that the *-o* in *hydro* is dropped.

The naming of the oxyacids is a little more complicated than that of the hydroacids. The nomenclature is based on the oxidation state of the central atom. Let's consider the common acids H_2SO_4, HNO_3, and H_3PO_4. In all three of these, the central atom is in its highest oxidation state.

$$\overset{6+}{H_2SO_4} \qquad \overset{5+}{HNO_3} \qquad \overset{5+}{H_3PO_4}$$

Remember that among the representative elements, the highest positive oxidation state is always equal to the group number. Now, to name these acids, one simply adds an -ic ending to the stem of the central atom and then follows this with the word acid. For our examples we have

H_2SO_4 sulfur*ic* acid
HNO_3 nitr*ic* acid
H_3PO_4 orthophosphor*ic* acid

Next we should look at acids such as H_2SO_3, HNO_2, and H_3PO_3 in which the central atom of the oxyacid is in its next highest oxidation state. Here we have

$$\overset{4+}{H_2SO_3} \qquad \overset{3+}{HNO_2} \qquad \overset{3+}{H_3PO_3}$$

In this case, the acids are named by adding an -ous ending to the stem of the central atom and then following with the word acid:

H_2SO_3 sulfur*ous* acid
HNO_2 nitr*ous* acid
H_3PO_3 orthophosphor*ous* acid

So we see that the higher oxidation state of the central atom gives us the -ic ending, whereas the lower oxidation state gives us the -ous ending.

In all of these examples, the central atom tends to show only two positive oxidation states, and this conveniently results in the -ic and the -ous endings. On the other hand, in the oxyhalogen acids, the halogen can show four positive oxidation states, $7+, 5+, 3+$, and $1+$. The two middle oxidation states are given the endings -ic and -ous:

$$\overset{5+}{HClO_3} \qquad \text{chlor}\textit{ic}\ \text{acid}$$

$$\overset{3+}{HClO_2} \qquad \text{chlor}\textit{ous}\ \text{acid}$$

The acid in which the halogen has the 7+ oxidation state results in a *per-ic* acid, and the 1+ oxidation state results in a *hypo-ous* acid:

$$\overset{7+}{H}ClO_4 \qquad \textit{perchloric} \text{ acid}$$
$$\overset{1+}{H}ClO \qquad \textit{hypochlorous} \text{ acid}$$

By recalling the discussion in Chapter 3, you should now begin to see the relationship between the ending in the name of a polyatomic anion and that of the corresponding acid: namely, we find an *-ate* ion results in an *-ic* acid, whereas an *-ite* ion results in an *-ous* acid. And, of course, it is also necessary to keep the prefixes such as *ortho-* and *meta-* as well as *per-* and *hypo-* straight.

———————— **EXERCISE** ————————

5.1 Here we have an easy one. Indicate whether each of the following is a hydroacid or an oxyacid.

a. HCN _____   **b.** HPO_3 _____

c. HN_3 _____   **d.** H_5IO_6 _____

e. H_2S _____   **f.** H_2CrO_4 _____

The following table summarizes the nomenclature of oxyacids. Particularly note the nomenclature of the oxychlorine anions and acids.

Polyatomic ion	Ion name	Oxyacid	Acid name
ClO_4^-	*perchlorate*	$HClO_4$	*perchloric*
MnO_4^-	*permanganate*	$HMnO_4$	*permanganic*
CO_3^{2-}	*carbonate*	H_2CO_3	*carbonic*
SO_4^{2-}	*sulfate*	H_2SO_4	*sulfuric*
NO_3^-	*nitrate*	HNO_3	*nitric*
PO_3^-	*metaphosphate*	HPO_3	*metaphosphoric*
PO_4^{3-}	*orthophosphate*	H_3PO_4	*orthophosphoric*
ClO_3^-	*chlorate*	$HClO_3$	*chloric*
SO_3^{2-}	*sulfite*	H_2SO_3	*sulfurous*
NO_2^-	*nitrite*	HNO_2	*nitrous*
ClO_2^-	*chlorite*	$HClO_2$	*chlorous*
ClO^-	*hypochlorite*	$HClO$	*hypochlorous*

With this background, you should now be able to name the common inorganic acids. So here is a chance to demonstrate how much you have learned.

_____ **EXERCISE** _____

5.2 Without looking at the summary, name these acids. If you must, use the periodic table.

a. HBr _____   **b.** H_2SO_3 _____

c. HNO_2 _____   **d.** H_2Se _____

e. $HClO_4$ _____   **f.** HPO_3 _____

g. H_2CrO_4 _____   **h.** HF _____

i. $HAsO_3$ _____   **j.** H_2SeO_4 _____

k. $HClO_2$ _____   **l.** H_6TeO_6 _____

m. HIO_3 _____   **n.** HCN _____

o. H_3PO_3 _____   **p.** HClO _____

Preparation of Brønsted-Lowry Acids. There are many acids, and some are rather difficult to prepare. Yet, there are a number of quite common acids that can be made by simple *type* reactions. These will be considered next, with special emphasis on specific "type" reactions.

(a) *Direct Combination Reactions.* Most of the hydroacids can be made by direct combination reactions. However, this may not be the safest or most convenient means for their preparation. For example, the direct preparation of the hydrohalogens is limited because H_2 and F_2 react explosively and H_2 and I_2 react very slowly at room temperature. In the latter case, if the temperature is raised to speed up the reaction, the HI dissociates back to H_2 and I_2. In preparing hydroacids containing the VIA elements, only H_2O can be made by direct combination. (Although not normally considered to be an acid, we will see that H_2O sometimes behaves as one.) Fortunately other methods exist for the preparation of the hydroacids.

————— **EXERCISE** —————

5.3 These should all be easy for you. Complete and balance the following reactions:

a. $H_2 + Cl_2 \longrightarrow$ _____

b. $H_2 + F_2 \longrightarrow$ _____

c. $H_2 + Br_2 \longrightarrow$ _____

d. $H_2 + I_2 \longrightarrow$ _____

(b) *Acid Anhydrides and Water.* Using a type of direct combination reaction (molecule + molecule), many of the oxyacids can be made by the reaction of a nonmetal oxide and water. In the majority of the oxyacids, and certainly in the more common ones, the oxidation state of the central atom will be one of the states we would predict from the electron configuration. For instance, note the oxidation state of the central atom in each of the following:

$$\overset{4+}{H_2SO_3} \quad \overset{6+}{H_2SO_4} \quad \overset{3+}{H_3PO_3} \quad \overset{5+}{H_3PO_4} \quad \overset{3+}{HNO_2} \quad \overset{5+}{HNO_3}$$

Remember from Chapter 1 that these oxidation states correspond to those expected from the electron configurations.

————— **EXERCISE** —————

5.4 For some more review, give the oxidation state of the underlined element and the predicted oxidation states based on electron configuration.

a. $H\underline{Mn}O_4$ _____

b. $H\underline{P}O_2$ _____

c. $H_2\underline{Se}O_4$ _____

d. $\underline{P}O_4{}^{3-}$ _____

e. $\underline{N}_2O_5$ _____

f. $\underline{C}O_2$ _____

g. $H_2\underline{S}_2O_7$ _____

h. $\underline{S}O_3{}^{2-}$ _____

i. $H\underline{Cl}O_4$ _____

j. $H_6\underline{Te}O_6$ _____

Now consider the reaction

$$\overset{4+}{SO_2} + H_2O \longrightarrow \overset{4+}{H_2SO_3}$$

There are three things that should be specifically noted about this reaction:

1. There is only one product—the acid.
2. The oxidation state of sulfur is 4+ on both sides of the equation. In fact, there are no changes in oxidation states.
3. 4+ is an oxidation state one would expect for sulfur from its electron configuration: $[Ne]3s^23p^4$.

When an acid is the only product of the reaction between an oxide and water, the oxide is said to be an *acid anhydride*. Anhydrides are compounds from which water has been removed. Thus, an acid anhydride is what is left when water is removed from an acid. When water is added to an acid anhydride, the reverse occurs. The corresponding acid, and nothing else, is obtained. In the reaction above, SO_2 is the acid anhydride of the acid H_2SO_3. As a general rule, if the oxidation state of the central element in a nonmetal oxide is one of those you would predict on the basis of the electron configuration, then the oxide is an acid anhydride and will react with water to give the corresponding acid.

-------- **EXERCISE** --------

5.5 Before we try specific reactions, see if you can tell which of the following are acid anhydrides. For each anhydride, give the corresponding acid and indicate the oxidation state of the central atom.

a. $\overset{6+}{SO_3}$ _yes_ $\underline{H_2SO_4}$ **b.** CO_2 _____ _____

c. NO_2 _____ _____   **d.** Na_2O _____ _____

e. Cl_2O_7 _____ _____   **f.** P_4O_6 _____ _____

g. ClO_2 _____ _____   **h.** SeO_2 _____ _____

i. N_2O_5 _____ _____   **j.** NO _____ _____

It is now a simple matter to predict the product of the reaction between an acid anhydride and water, but let's try one more to be sure you've got it. Consider the reaction

$$P_4O_6 + H_2O \longrightarrow$$

First note that P_4O_6 is an oxide of a nonmetal. There is, therefore, a good chance that it is an acid anhydride. Next, note that the oxidation state of phosphorus in this compound is $3+$. This is an expected oxidation state of phosphorus, and we can now feel confident that P_4O_6 is an acid anhydride. The product, then, should be the oxyacid and nothing else, with the oxidation state of phosphorus in the acid also being $3+$:

$$\overset{3+}{P_4O_6} + H_2O \longrightarrow \overset{3+}{H_3PO_3}$$

or, if we balance the equation

$$3\ P_4O_6 + 18\ H_2O \longrightarrow 12\ H_3PO_3$$

Rather than H_3PO_3, you might expect to obtain the *meta-* form of the acid, HPO_2. This is a particularly interesting molecule. It appears to be metaphosphorous acid, and, in fact, it does exist and is often referred to by this name. However, HPO_2 is really not an acid. The hydrogen atom is bonded directly to the phosphorus atom rather than to an oxygen atom. Thus, it is not an acidic hydrogen. Because of this, the phosphite ion, PO_2^- or PO_3^{3-}, does not exist.

EXERCISE

5.6 Now you should have it. Complete the following reactions.

a. $CO_2 + H_2O \longrightarrow$ _____   **b.** $I_2O_5 + H_2O \longrightarrow$ _____

c. $SO_2 + H_2O \longrightarrow$ _____   **d.** $SeO_2 + H_2O \longrightarrow$ _____

e. $P_4O_6 + H_2O \longrightarrow$ _____   **f.** $Cl_2O_7 + H_2O \longrightarrow$ _____

g. $N_2O_5 + H_2O \longrightarrow$ _____   **h.** $SO_3 + H_2O \longrightarrow$ _____

i. $P_4O_{10} + H_2O \longrightarrow$ _____   **j.** $As_2O_3 + H_2O \longrightarrow$ _____

The reaction of an acid anhydride with water can be used to make an oxyacid, but it isn't commonly done. However, several of the examples shown in Exercise 5.6 do play an important role in our lives. For example, acid rain results from the reactions of SO_2 and SO_3 with H_2O in the atmosphere, and in a lighter vein, it is CO_2 from the equilibrium

$$CO_2 + H_2O \rightleftharpoons H_2CO_3$$

that makes the bubbles in carbonated drinks such as colas, beer, and champagne.

(c) *Salt and a Less Volatile Acid.* Sulfuric acid (H_2SO_4) is a viscous liquid having a boiling point of roughly 340°C. These properties allow us to make more volatile acids by distilling the more volatile acid from the reaction mixture of H_2SO_4 and a salt. Thus we have

$$CaF_2(s) + H_2SO_4(l) \xrightarrow{\Delta} CaSO_4(s) + 2\ HF(g)$$

This approach is used for the commercial preparation of a number of protonic acids, most notably HF and HCl. However, H_2SO_4 cannot be used to make HBr and HI. The H_2SO_4 oxidizes Br^- and I^- to Br_2 and I_2, respectively. To avoid this problem, one can use another nonvolatile acid. Phosphoric acid (H_3PO_4) is ordinarily used for the preparation of HBr and HI. In fact, H_2SO_4 and H_3PO_4 are probably the only acids you will run across that are used in this manner. Now note that, in principle, all that is needed to produce a given acid is a salt containing the anion of the desired acid and a less volatile acid. For instance, if one wishes to make HNO_3, a salt containing the NO_3^- ion is chosen. Interestingly, prior to 1914, the commercial production of HNO_3 was based on the reaction

$$2\ NaNO_3(s) + H_2SO_4(l) \xrightarrow{\Delta} Na_2SO_4(s) + 2\ HNO_3(g)$$

A useful variation of this approach takes advantage of the fact that some salts such as $BaSO_4$ are insoluble in an aqueous solution. One then chooses the appropriate reactants for a precipitation reaction such that the salt precipitates from the solution, leaving the desired acid. For instance, consider

$$Ba(ClO_3)_2(aq) + H_2SO_4(aq) \longrightarrow BaSO_4(s) + 2\ HClO_3(aq)$$

We, of course, are not limited to barium salts and H_2SO_4. It is only necessary that an insoluble salt be formed.

In principle, both of these approaches work well. However, it turns out that they can be efficiently applied to the preparations of only a few acids.

––––––––– **EXERCISE** –––––––––

5.7 Now for some practice. Complete the following.

a. $NaNO_3(s) + H_2SO_4 \xrightarrow{\Delta}$ _____

b. $NaCN(s) + H_2SO_4 \xrightarrow{\Delta}$ _____

c. $Ba(ClO_2)_2(aq) + H_2SO_4(aq) \longrightarrow$ _____

d. $NaI(s) + H_3PO_4 \xrightarrow{\Delta}$ _____

e. $KBr(s) + H_3PO_4 \xrightarrow{\Delta}$ _____

f. $NaCl(s) + H_2SO_4 \xrightarrow{\Delta}$ _____

g. $Ba(NO_3)_2(aq) + H_2SO_4(aq) \longrightarrow$ _____

h. $KF(s) + H_2SO_4 \xrightarrow{\Delta}$ _____

i. $Ba(ClO_4)_2(aq) + H_2SO_4(aq) \longrightarrow$ _____

j. $NaBr(s) + H_2SO_4 \xrightarrow{\Delta}$ _____

(d) *Phosphorus Halides and Oxyhalides with Water.* A limited group of acids can be made by the reaction of a phosphorus halide or oxyhalide with water. These are compounds such as PI_3, PCl_5, PBr_3, and $POCl_3$. Note that the oxidation state of phosphorus is either $3+$ or $5+$, states one would predict from the electron configuration of phosphorus: $[Ne]3s^2 3p^3$. These compounds react with water to give the hydrohalic acid, HX, and the oxyphosphorous acid in which phosphorus has the same oxidation state as it does in the reactant. Thus we have

$$\overset{3+}{PCl_3} + 3\ H_2O \longrightarrow 3\ HCl + \overset{3+}{H_3PO_3}$$

and
$$\overset{5+}{POCl_3} + 3\ H_2O \longrightarrow 3\ HCl + \overset{5+}{H_3PO_4}$$

The HCl in these reactions can then be obtained by heating the resultant solution. PBr_3 and PI_3 are actually used for the commercial preparation of HBr and HI.

_____ **EXERCISES** _____

5.8 Complete the following reactions.

a. $PI_3 + H_2O \longrightarrow$ _____

b. $POBr_3 + H_2O \longrightarrow$ _____

c. $PCl_5 + H_2O \longrightarrow$ _____

d. $PCl_3 + H_2O \longrightarrow$ _____

5.9 Give three reactions for preparing HCl.

a. _____

b. _____

c. _____

5.10 Give two reactions for preparing HIO_3.

a. _____

b. _____

5.11 Give two reactions for preparing H_3PO_4.

a. _____

b. _____

5.12 Give two reactions for preparing HNO_2.

a. _____

b. _____

BASES

Just as with acids, bases have also gone through a long historical development with a number of useful definitions having been made. But again we find that the protonic definition is the one most commonly used. Accordingly, a base is defined as any substance capable of accepting a proton. Thus, whereas an acid is a proton donor, a base is a *proton acceptor*.

Before the protonic definition, a base was considered to be a hydroxyl compound that yields OH^- ions in aqueous solution. Basic properties were, therefore, defined in terms of the OH^- ion. Normally, hydroxyl bases are metal hydroxides such as NaOH, although, except for the alkali metal hydroxides, most are relatively insoluble in water.

Preparation of Hydroxyl Bases. Many of the metal hydroxides can be made by the reaction of the corresponding metal oxide with water. Thus, there exist basic anhydrides as well as acid anhydrides, and the reactions of both types of compounds with water are very much alike. But, whereas an acid anhydride is usually a nonmetal oxide, a basic anhydride will be a metal oxide. Two examples of the reaction of a metal oxide with water are

$$Na_2O + H_2O \longrightarrow 2\ NaOH$$

$$CaO + H_2O \longrightarrow Ca(OH)_2$$

It is also possible to decompose a metal hydroxide by heating it. This type of reaction is an example of *thermal decomposition*. If there is no redox reaction—that is, if no changes in oxidation states occur—then the only possible products will be the metal oxide and water. Consider the following examples:

$$Mg(OH)_2(s) \xrightarrow{\ \Delta\ } MgO(s) + H_2O(g)$$

$$Ca(OH)_2(s) \xrightarrow{\ \Delta\ } CaO(s) + H_2O(g)$$

$$2\ Al(OH)_3(s) \xrightarrow{\ \Delta\ } Al_2O_3(s) + 3\ H_2O(g)$$

——————— **EXERCISES** ———————

5.13 As a review, for each of the following, identify the type of compound (oxide, peroxide, or superoxide) and state whether or not it is a basic anhydride.

a. CsO_2 superoxide   no   **b.** BaO _____ _____

c. Na_2O_2 _____ _____   **d.** La_2O_3 _____ _____

e. N_2O_3 _____ _____   **f.** NaO_2 _____ _____

g. K_2O _____ _____   **h.** SrO_2 _____ _____

i. RbO_2 _____ _____   **j.** Ag_2O _____ _____

5.14 Now let's see if you can use this information. Complete the following reactions.

a. $BaO + H_2O \longrightarrow$ _____

b. $La_2O_3 + H_2O \longrightarrow$ _____

c. $Zn(OH)_2 \xrightarrow{\ \Delta\ }$ _____

d. $Na_2O + H_2O \longrightarrow$ _____

e. $KOH \xrightarrow{\ \Delta\ }$ _____

f. $Al_2O_3 + H_2O \longrightarrow$ _____

g. $Ag_2O + H_2O \longrightarrow$ _____

h. $CaO + H_2O \longrightarrow$ _____

i. $Sr(OH)_2 \xrightarrow{\Delta}$ _____

j. $Cs_2O + H_2O \longrightarrow$ _____

NEUTRALIZATION

In an aqueous solution, water dissociates slightly to give a very small concentration of H_3O^+ and OH^- ions according to the equation

$$2\,H_2O \rightleftharpoons H_3O^+(aq) + OH^-(aq)$$

and, if only H_2O is involved, the number of H_3O^+ ions must equal the number of OH^- ions. Under such circumstances, the solution is neutral. However, if an acid is added, there will be an excess of H_3O^+ ions, and the solution will be acidic. On the other hand, if a hydroxyl base is added, there will be an excess of OH^- ions, and the solution will be basic. Now, if an acid is added to a solution containing an excess of OH^- ions, the acid can neutralize the base. If the reverse occurs—that is, a hydroxyl base is added to a solution containing an excess of H_3O^+ ions—the base can neutralize the acid. This is an important type of reaction between acids and bases and is referred to as *neutralization*. For a protonic acid and a hydroxyl base, neutralization is simply the combination of H^+ ions and OH^- ions to form H_2O:

$$H^+(aq) + OH^-(aq) \rightleftharpoons H_2O$$

Or, we might write the reaction more accurately as

$$H_3O^+(aq) + OH^-(aq) \rightleftharpoons 2\,H_2O$$

As a specific example of a neutralization reaction, we can write

$$NaOH + HCl \xrightarrow{(H_2O)} NaCl(aq) + H_2O$$

But it should be recognized that in an aqueous solution,

$$NaOH \xrightarrow{(H_2O)} Na^+(aq) + OH^-(aq)$$

and $\qquad\qquad HCl \xrightarrow{(H_2O)} H_3O^+(aq) + Cl^-(aq)$

Since the NaCl remains in solution as Na^+ and Cl^-, the only reaction that occurs is simply

$$H_3O^+(aq) + OH^-(aq) \rightleftharpoons 2\ H_2O$$

_____ **EXERCISE** _____

5.15   Here's your chance to try a new *type* reaction, the reaction of a hydroxyl base with a protonic acid. We'll work the first one for you. Assume all reactions are in aqueous solution.

a. $NaOH + HBr \longrightarrow$ $\underline{NaBr + H_2O}$

b. $KOH + HCl \longrightarrow$ _____

c. $CsOH + HI \longrightarrow$ _____

d. $AgOH + HCl \longrightarrow$ _____

e. $RbOH + HBr \longrightarrow$ _____

_____

With a polyprotic acid—that is, an acid such as H_2SO_4 or H_3PO_4 which has more than one acidic hydrogen atom—all of the acidic hydrogens will be replaced if sufficient base is present. Similarly, all of the OH^- groups will be replaced in a polyhydroxyl base if a sufficient concentration of protonic acid is present. For complete neutralization in an aqueous solution, then, we have as examples

$$H_3PO_4 + 3\ NaOH \longrightarrow Na_3PO_4 + 3\ H_2O$$

$$2\ HCl + Ba(OH)_2 \longrightarrow BaCl_2 + 2\ H_2O$$

Be sure to keep in mind that there are no changes in oxidation states in an acid-base reaction.

_____ **EXERCISE** _____

5.16   Complete and balance the following reactions. These should be a little more of a challenge than those in the last exercise. Assume complete neutralization and aqueous solutions. Again, the first one is worked for you.

a. $3\ HCl + Al(OH)_3 \longrightarrow$ $\underline{AlCl_3 + 3\ H_2O}$

b. $2\ HBr + Ca(OH)_2 \longrightarrow$ _____

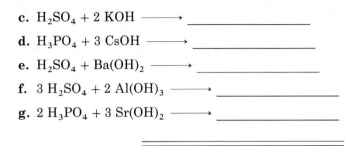

c. $H_2SO_4 + 2\ KOH \longrightarrow$ _____

d. $H_3PO_4 + 3\ CsOH \longrightarrow$ _____

e. $H_2SO_4 + Ba(OH)_2 \longrightarrow$ _____

f. $3\ H_2SO_4 + 2\ Al(OH)_3 \longrightarrow$ _____

g. $2\ H_3PO_4 + 3\ Sr(OH)_2 \longrightarrow$ _____

BRØNSTED-LOWRY ACID-BASE REACTIONS

We have already defined acids and bases in terms of the Brønsted-Lowry definition. Recall that acids are proton donors and bases are proton acceptors. First, let's look at what happens when an acid is dissolved in a solvent such as water. In most cases, we find that an equilibrium results. In general, this can be expressed as

$$HA(aq) + H_2O \rightleftharpoons H_3O^+(aq) + A^-(aq)$$

where A represents any anion. However, if the acid is a strong acid, it is assumed to be 100% dissociated, giving

$$HA + H_2O \xrightarrow{100\%} H_3O^+(aq) + A^-(aq)$$

As it turns out, there are very few acids that can be categorized as strong acids. In aqueous solution the common ones are HI, HBr, HCl, $HClO_4$, $HClO_3$, HNO_3, and H_2SO_4 (only the first hydrogen in H_2SO_4 is 100% ionized). The remaining common acids can be considered to be weak acids, and in water an equilibrium exists as shown in the first equation above. Let's look at a specific example:

$$HF(aq) + H_2O \rightleftharpoons H_3O^+(aq) + F^-(aq)$$

Now, recalling that a Brønsted-Lowry acid donates a proton and a Brønsted-Lowry base accepts a proton, it can be seen that we are actually dealing here with a protonic acid-base reaction. Hydrofluoric acid (HF) is donating a proton to the H_2O molecule and is therefore an acid. At the same time, the H_2O accepts the proton and thereby functions as a base.

Conjugate Acids and Bases. The reaction above of a weak acid with water was shown with a double arrow indicating that an equilibrium exists. This, of course, means that both forward and reverse reactions are occurring, and the reverse reaction is an acid-base reaction as well. The H_3O^+ donates a proton and thereby acts as an acid, while the F^- accepts a proton and acts as a base. To distinguish between these two acid-base pairs, we refer to those species on the right side of the equation as the *conjugate* acid and the *conjugate* base and to those on the left side as simply the acid and the base. Let's look at one more example.

$$H_3PO_4(aq) + H_2O \rightleftharpoons H_3O^+(aq) + H_2PO_4^-(aq)$$

   acid   base   conjugate   conjugate
                                acid   base

_____   **EXERCISE**   _____

5.17   Before going on, be sure you can identify the strong acids in aqueous solution. Circle those that act as strong acids.

a. (HCl)   **b.** H_3PO_4

c. H_2S **d.** HNO_2

e. H_2SO_3 **f.** $HMnO_4$

g. HCN   **h.** $HClO_2$

i. HI   **j.** $HClO_4$

Ammonia as a Base.   An interesting aspect of the Brønsted-Lowry acid-base behavior is that a given substance can act as either an acid or a base, depending on the reaction. For instance, water behaves as a base when reacting with HF, but with NH_3 we can see that it behaves as an acid:

$$H_2O + NH_3(aq) \rightleftharpoons NH_4^+(aq) + OH^-(aq)$$

Here H_2O donates a proton to NH_3 and is therefore acidic. Furthermore, the fact that H_2O does donate the proton to NH_3 tells us that NH_3 is a stronger base than H_2O. We might, then, logically conclude that the reaction of a protonic acid with ammonia will be similar to the reaction of a protonic acid with water, and this is the case, as can be seen from the reaction

$$HCl(aq) + NH_3(aq) \longrightarrow NH_4^+(aq) + Cl^-(aq)$$

5.18 Now for the next step. Complete the following reactions, and identify the base and the conjugate acid.

	Base	Conjugate acid
a. $HNO_2 + H_2O \rightleftharpoons$ _____	_____	H_3O^+
b. $HCN + H_2O \rightleftharpoons$ _____	_____	_____
c. $HClO_2 + NH_3 \rightleftharpoons$ _____	_____	_____
d. $H_2S + H_2O \rightleftharpoons$ _____	_____	_____
e. $HCN + NH_3 \rightleftharpoons$ _____	_____	_____

TYPES OF BRØNSTED-LOWRY BASES

In general, the identification of a protonic acid is relatively simple: if a substance (Brønsted-Lowry acid) donates a proton, it obviously must contain a hydrogen atom. No such generalization, however, can be made for a protonic base. There are many molecules and ions that can accept a proton and act as a Brønsted-Lowry base. Thus far, we have seen the two common types: molecules such as H_2O and anions such as F^-. However, in our example using the F^- ion, the anion appeared as a conjugate base. Let's first consider the anions.

As a general rule, anions of weak acids, anions such as NO_2^-, ClO^-, $H_2PO_3^-$, HPO_3^{2-}, F^-, etc., will act as bases in the presence of H_2O. Water will then have to act as an acid, as can be seen in the following reaction with the CN^- ion:

$$CN^-(aq) + H_2O \rightleftharpoons HCN(aq) + OH^-(aq)$$

With the polyprotic acid H_2SO_3, both the SO_3^{2-} ion and the HSO_3^- ion are Brønsted-Lowry bases in water, as we can see here:

$$SO_3^{2-}(aq) + H_2O \rightleftharpoons HSO_3^-(aq) + OH^-(aq)$$

$$HSO_3^-(aq) + H_2O \rightleftharpoons H_2SO_3(aq) + OH^-(aq)$$

This same situation exists with all of the common polyprotic acids except H_2SO_4. With H_2SO_4 only the SO_4^{2-} ion is a base, reacting with water to give

$$SO_4^{2-}(aq) + H_2O \rightleftharpoons HSO_4^-(aq) + OH^-(aq)$$

Recall that HSO_4^- is the anion formed from the ionization of the first hydrogen in H_2SO_4, which is a strong acid. Actually, H_2SO_4 is the only example of this problem. All other polyprotic acids are weak acids.

As you have probably realized, you don't have to be concerned here with the anions of strong acids. The fact that these acids ionize 100% (that's why they are considered to be strong acids) tells us that their anions have no particular affinity for protons. Thus, they don't act as proton acceptors. Finally, it is easy to recognize the anions of the weak acids. You just have to memorize the seven common strong acids and remember that the anions of all the rest are anions of weak acids.

You have also seen that molecules such as H_2O and NH_3 can act as bases when they come in contact with substances that are more acidic than they are. This includes essentially all of the common acids. Just remember that when any protonic acid reacts with H_2O or NH_3, the H_2O or the NH_3 will in essentially every instance act as a Brønsted-Lowry base by accepting a proton.

There are many other molecules besides H_2O and NH_3 that will show similar basic properties. In fact, compounds containing oxygen and nitrogen frequently show this behavior. As one specific example, we might consider methyl alcohol, CH_3OH. Its behavior is very similar to that of water, and the reaction of HCl with each clearly shows this similarity:

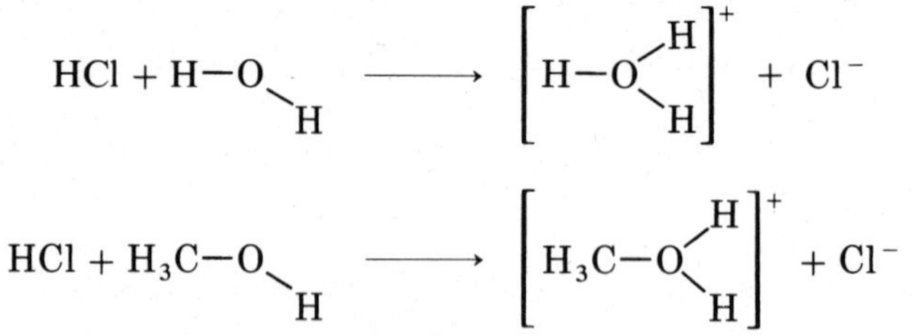

$$HCl + H-O\begin{smallmatrix}\\H\end{smallmatrix} \longrightarrow \left[H-O\begin{smallmatrix}H\\H\end{smallmatrix}\right]^+ + Cl^-$$

$$HCl + H_3C-O\begin{smallmatrix}\\H\end{smallmatrix} \longrightarrow \left[H_3C-O\begin{smallmatrix}H\\H\end{smallmatrix}\right]^+ + Cl^-$$

_____ **EXERCISE** _____

5.19 It is now time for the final test for this chapter. See if you can complete the following reactions and identify the acid and the conjugate base. For polyprotic acids, complete only the ionization of one proton.

	Acid	Conjugate base

a. $ClO_2^- + H_2O \rightleftharpoons$ _____ _____ _____

b. $H_2PO_3^- + NH_3 \rightleftharpoons$ _____ _____ _____

c. $NO_2^- + H_2O \rightleftharpoons$ _____ _____ _____

d. $H_2CO_3 + H_2O \rightleftharpoons$ _____ _____ _____

e. $HSO_3^- + H_2O \rightleftharpoons$ _____ _____ _____

f. $HNO_2 + NH_3 \rightleftharpoons$ _____ _____ _____

g. $H_3PO_3 + H_2O \rightleftharpoons$ _____ _____ _____

h. $H_2PO_3^- + H_2O \rightleftharpoons$ _____ _____ _____

i. $HCN + NH_3 \rightleftharpoons$ _____ _____ _____

j. $CN^- + NH_3 \rightleftharpoons$ _____ _____ _____

6

Acid-Base II: Hydrolysis

After finishing Chapter 5, you might conclude that acid-base behavior can easily be rationalized. Substances that donate protons are acids and substances that accept protons are bases. It turns out, however, that *salts* can also exhibit acidic or basic characteristics. As one might expect, it is not nearly so easy to explain such behavior in salts as it is to explain it in substances which already contain hydrogen atoms or hydroxide ions.

Let's now turn to what seems like a simple question but, in fact, is really quite deceptive. What happens when a salt is put in water? Of course, you might say that it dissolves! But we should be interested in more than that. For instance, let's consider what happens to the pH[1] of each solution when different salts are dissolved. After all, you might say, "salts can be considered to be the neutralization products of acid-base reactions. Thus, an aqueous solution of a salt should be neutral—that is, have a pH equal to 7." The actual results, however, may be quite unexpected. There are three distinct groups of salt solutions that we will discuss. Solutions of Type I do, indeed, have a pH of 7. Solutions of Type II have pH values greater than 7 (basic), while solutions of Type III have pH

---

[1] Recall that pH is a measure of acidity in solution: $pH = -\log [H_3O^+]$. A solution that has a pH equal to seven is neutral. A solution with a pH less than seven is acidic, and one with a pH greater than seven is basic.

values less than 7 (acidic). Representative examples are

Type I	Type II	Type III
NaCl	KF	NH_4Cl
BaI_2	$NaOAc^2$	CuI_2
KBr	RbCN	$Zn(NO_3)_2$

These results will be explained by first analyzing each of the groups to see what the members have in common. Later on in the chapter we shall consider the reasons for this behavior.

Before proceeding with this discussion, let's list the hydroxyl bases that are commonly considered to be strong. These are all of the group **IA** hydroxides and three of the group **IIA** hydroxides.

Group **IA**	Group **IIA**
LiOH	
NaOH	
KOH	$Ca(OH)_2$
RbOH	$Sr(OH)_2$
CsOH	$Ba(OH)_2$

_____ **EXERCISE** _____

6.1   Review! List the common strong acids.

_____

_____

_____

===

Next recall that a salt can be considered to be the neutralization product of a protonic acid and a hydroxyl base. In general, we can say

$$acid + base \longrightarrow salt + H_2O$$

and taking a specific example,

cation from base

$$HCl(aq) + NaOH(aq) \longrightarrow NaCl(aq) + H_2O$$

anion from acid

---

[2] OAc^- is a shorthand representation for the acetate anion CH_3COO^-, which is the anion of acetic acid, CH_3COOH. The hydrogen attached to an oxygen atom is the acidic hydrogen.

Note that the cation of the salt is always derived from the base while the anion of the salt is derived from the acid.

ACIDIC, BASIC, AND NEUTRAL SALTS[3]: THE BEHAVIOR

Neutral Salts.   Recalling the examples we gave for Type I salts, we can see that the common factor among them is that they are all salts of strong acids and strong bases:

$$NaOH(aq) + HCl(aq) \longrightarrow \underline{NaCl}(aq) + H_2O$$

$$Ba(OH)_2(aq) + 2\ HI(aq) \longrightarrow \underline{BaI_2}(aq) + 2\ H_2O$$

$$KOH(aq) + HBr(aq) \longrightarrow \underline{KBr}(aq) + H_2O$$

The important observation is that *salts that are formed from strong acids and strong bases give neutral solutions.*

Basic Salts.   Consider now the salts whose solutions have pH values greater than 7 (basic). These salts are called basic salts and are formed from the reactions of strong bases with weak acids. As a mnemonic device remember that strong dominates weak; thus *salts that are formed from strong bases and weak acids give basic solutions.*

For example, look at the formation of Type II salts:

$$KOH(aq) + HF(aq) \longrightarrow \underline{KF}(aq) + H_2O$$

$$NaOH(aq) + HOAc(aq) \longrightarrow \underline{NaOAc}(aq) + H_2O$$

$$RbOH(aq) + HCN(aq) \longrightarrow \underline{RbCN}(aq) + H_2O$$

_____   **EXERCISE**   _____

6.2   Again:

a. KF is derived from the strong base _____ and the _____ acid HF.

---

[3] This is not a common terminology, but it is very convenient. By *acidic, basic,* and *neutral* salts we mean salts that give acidic, basic, and neutral aqueous solutions, respectively. The names acidic salt and acid salt should not be confused. An acid salt is a salt such as $NaHSO_4$ or NaH_2PO_4 which contains one or more ionizable hydrogen atoms. A salt containing no ionizable hydrogen atoms (for instance, Na_2SO_4 or Na_3PO_4) is sometimes called a *neutral* salt, but a more appropriate name is *normal* salt. We will refer to a neutral salt as one whose aqueous solution has a pH of 7.

b. RbCN is derived from the _____ base RbOH and the weak acid

_____ .

===============================

Acidic Salts. The salts that have pH values less than 7 (acidic), called acidic salts, then are formed from the reactions of strong acids with weak bases. Again follow the mnemonic device: *salts that are formed from strong acids and weak bases give acidic solutions.*

Examples of the formation of Type III salts then are

$$NH_3(aq) + HCl(aq) \longrightarrow \underline{NH_4Cl}(aq)$$

$$Cu(OH)_2(aq) + 2\ HI(aq) \longrightarrow \underline{CuI_2}(aq) + 2\ H_2O$$

$$Zn(OH)_2(aq) + 2\ HNO_3(aq) \longrightarrow \underline{Zn(NO_3)_2}(aq) + 2\ H_2O$$

_____ **EXERCISES** _____

6.3   Now for a quick review. Give the formulas of the acid and the base that will form each of the following salts.

Salt	Acid	Base
a. NaI	HI	NaOH
b. CsF	___	___
c. CaBr$_2$	___	___
d. NH$_4$Br	___	___
e. Fe(ClO$_2$)$_2$	___	___
f. Na$_3$PO$_4$	___	___
g. Al$_2$(SO$_4$)$_3$	___	___

6.4   Next consider the important part of this discussion! State whether the aqueous solutions of the following salts are acidic, basic, or neutral.

a. CuBr$_2$ _____   **b.** KOAc _____

c. BaCl$_2$ _____   **d.** NH$_4$I _____

e. CsCl _____   **f.** Na$_3$PO$_4$ _____

g. $RbNO_3$ _____   **h.** $FeCl_3$ _____

i. $Zn(ClO_3)_2$ _____   **j.** KCN _____

k. $NaClO_4$ _____

=======================

ACIDIC, BASIC, AND NEUTRAL SALTS: THE RATIONALE

Neutral Salts. Let's now return to the original question and discuss why we get the previously mentioned experimental results. Remember that the salts of strong acids and strong bases do not change the pH of an aqueous solution. Thus there is no net excess of H_3O^+ or OH^- ions. The solution contains only the cation and the anion of the salt in water; for example,

$$NaCl(s) + H_2O \longrightarrow Na^+(aq) + Cl^-(aq)$$

$$KI(s) + H_2O \longrightarrow K^+(aq) + I^-(aq)$$

Such ions are called *hydrated* ions, meaning that the ions are surrounded by some numbers of water molecules. These hydrated ions are also called spectator ions because they do not react to form new species.

Our understanding of the meaning of a strong acid and a strong base can help us to see why no reaction occurs. Recall that a strong acid is one that is 100% ionized. Thus, for example,

$$HCl(aq) + H_2O \xrightarrow{100\%} H_3O^+(aq) + Cl^-(aq)$$

From the fact that the HCl is 100% ionized, we can conclude that the Cl^- ion cannot take a H^+ ion from H_3O^+. Otherwise, there would be an equilibrium such as is observed with the weak acid HF, namely

$$HF(aq) + H_2O \rightleftharpoons H_3O^+(aq) + F^-(aq)$$

Here it can be seen that the F^- ion can and does extract a H^+ ion from H_3O^+. This is really telling us that the F^- ion is a much better protonic base than the Cl^- ion. Now, if the Cl^- ion cannot take a H^+ ion from H_3O^+, it certainly can't take one from H_2O. Consequently, only hydration occurs:

$$Cl^- + H_2O \longrightarrow Cl^-(aq)$$

Quite commonly, this is expressed simply as

$$Cl^- + H_2O \longrightarrow N.R.$$

The important point is that if Cl^- ions are somehow placed in water, they will *not* react with the H_2O molecules in such a way as to form H_3O^+ or OH^- ions.

The same type of argument can be given for the cation from a strong hydroxyl base. Consider that NaOH is highly soluble in H_2O and ionizes 100%:

$$NaOH(s) \xrightarrow[\text{(H}_2\text{O)}]{100\%} Na^+(aq) + OH^-(aq)$$

Since the Na^+ ion has no tendency to combine with the OH^- ion to reform NaOH(s) or even NaOH(aq), one would not expect it to extract an OH^- from H_2O. As with the Cl^- ion, we have

$$Na^+ + H_2O \longrightarrow Na^+(aq)$$

or $\qquad\qquad\quad Na^+ + H_2O \longrightarrow N.R.$

This nonreactivity with water is observed with cations of all strong bases and anions of all strong acids.

Now, the question is *how do we get these ions in solution?* One way, and the way we are considering in this chapter, is by dissolving a salt in water. So,

$$NaCl(s) + H_2O \longrightarrow Na^+(aq) + Cl^-(aq)$$

and all we have are hydrated ions. Thus it now can be seen why an aqueous solution of a salt of a strong acid and a strong base has the same pH as does water.

<hr>   **EXERCISES**   <hr>

6.5 Let's now determine the products of some reactions between strong acids and strong bases.

a. $HCl(aq) + KOH(aq) \longrightarrow$ _____ + _____

b. $HNO_3(aq) + NaOH(aq) \longrightarrow$ _____ + _____

c. $2\ HI(aq) + Ca(OH)_2(aq) \longrightarrow$ _____ + _____

d. $2\ HBr(aq) + Ba(OH)_2(aq) \longrightarrow$ _____ + _____

6.6 Some more review! Write the ionic products of the following reactions. Don't forget to denote the aqueous phase. (You get a little help with the first one.)

a. $KCl(s) + H_2O \longrightarrow$ _____(aq)_____ $+$ ___$Cl^-(aq)$___

b. $NaNO_3(s) + H_2O \longrightarrow$ _____ $+$ _____

c. $CaI_2(s) + H_2O \longrightarrow$ _____ $+$ _____

d. $BaBr_2(s) + H_2O \longrightarrow$ _____ $+$ _____

e. $KClO_4(s) + H_2O \longrightarrow$ _____ $+$ _____

6.7 And even more review! Answer acidic, basic, or neutral.

a. Salts of strong acids and strong bases are _____ in aqueous solution.

b. Salts of strong bases and weak acids are _____ in aqueous solution.

c. Salts of strong acids and weak bases are _____ in aqueous solution.

Basic Salts. It wasn't too difficult to see why the neutral salts are, in fact, neutral. But now we face a somewhat greater challenge: how can we rationalize (that is, write equations for) the behavior of acidic and basic salts in water? The basic salts will be considered first, with KF as a representative example. Remember that

$$KF(s) + H_2O \longrightarrow \text{basic solution}$$

Although this is a correct statement, we should not be satisfied with merely recognizing that the solution is basic. We want to write an equation with all of the chemical species represented. Since the solution is basic (pH greater than 7), the concentration of OH^- ions must be greater than that of the H_3O^+ ions. Our goal is to find the source of the excess OH^- ions.

First let's recall that salts are 100% ionized in aqueous solution, so

$$KF(aq) + H_2O \xrightarrow{\ 100\%\ } K^+(aq) + F^-(aq)$$

There is no way for OH^- ions to be generated here! However, the F^- ion is the anion of a weak acid. In Chapter 5 we learned that in the reaction of an anion of a weak acid with H_2O, the anion acts as a protonic base. Using the F^- ion as our example, we get

$$F^-(aq) + H_2O \rightleftharpoons HF(aq) + OH^-(aq)$$

_____ **EXERCISE** _____

6.8 Note that the above is a protonic acid-base reaction in which _____ is the acid and _____ is the base.

This is an example of a *hydrolysis* reaction. In general, hydrolysis is a chemical reaction in which the water molecule is split. Depending on the reactants, there are a number of paths such a reaction can take. For the hydrolysis reaction between an ionic species and H_2O there will be a change in pH. In the example above, OH^- ions are formed, resulting in an increase in the pH. In fact, anions of weak acids will always react with H_2O to form OH^- ions and, therefore, increase the pH. At the same time, again note that the cations of strong bases do *not* hydrolyze. Rather, they will simply be hydrated, and thus will not change the pH.

_____ **EXERCISE** _____

6.9 Circle the ions which undergo hydrolysis. Remember that anions of weak acids hydrolyze. For example, F^- should be circled because it is the anion of the weak acid HF. Cl^- should not be circled because it is the anion of the strong acid HCl.

a. F^- **b.** Cl^-

c. CN^- **d.** K^+

e. Ba^{2+} **f.** I^-

g. Ca^{2+} **h.** OAc^-

i. NO_3^- **j.** Na^+

k. ClO_4^- **l.** Sr^{2+}

m. Cs^+ **n.** NO_2^-

o. Rb^+

To be certain that you understand the principles, let's consider in detail what happens when NaCN is dissolved in H_2O. In review, what do we know about NaCN?

a. It is a salt that is formed from the weak acid HCN and the strong base NaOH:

$$HCN(aq) + NaOH(aq) \longrightarrow NaCN(aq) + H_2O$$

b. Salts are 100% ionized in aqueous solution, so

$$NaCN(s) + H_2O \longrightarrow Na^+(aq) + CN^-(aq)$$

c. NaCN is a basic salt: it gives aqueous solutions that have a pH greater than 7.

EXERCISES

6.10   We know that NaCN is a basic salt because it is a salt of

a _____ acid and a _____ base. The Na^+ ion undergoes

_____ , and the CN^- ion undergoes _____ .

6.11   Write the equation for the hydration of the Na^+ ion.

$$Na^+ + H_2O \longrightarrow \underline{\hspace{3cm}}$$

6.12   O.K. Now let's write the products of the hydrolysis reaction of the CN^- ion.

$$CN^-(aq) + H_2O \rightleftharpoons \underline{\hspace{2cm}} + \underline{\hspace{2cm}}$$

6.13   Now complete the equations for the entire process of dissolving the basic salt NaCN in water.

a. the dissolving of the salt,

$$NaCN(s) + H_2O \longrightarrow \underline{\hspace{3cm}}$$

b. the hydration of the Na^+ ion,

$$Na^+ + H_2O \longrightarrow \underline{\hspace{3cm}}$$

c. the hydrolysis of the CN^- ion,

$$CN^-(aq) + H_2O \rightleftharpoons \underline{\hspace{6cm}}$$

$$\underline{\hspace{9cm}}$$

We can write the total reaction for the dissolution of NaCN in water as

$$Na^+(aq) + CN^-(aq) + H_2O \rightleftharpoons Na^+(aq) + HCN(aq) + OH^-(aq)$$

But the Na^+ ions, being spectator ions, cancel out, leaving the net ionic equation

$$CN^-(aq) + H_2O \rightleftharpoons HCN(aq) + OH^-(aq)$$

which is simply the hydrolysis equation for the CN^- ion! All hydrolysis reactions of basic salts, salts which are derived from a strong base and a weak acid, can be written in this same way.

_____ **EXERCISES** _____

6.14   Again and again! How do you identify a basic salt? It is a salt made from a strong _____ and a weak _____ . The cation of the salt comes from the _____ and undergoes _____ in aqueous solution.

6.15   Circle the basic salts among the following.

$$NaBr \quad KF \quad NaOAc \quad Ba(NO_3)_2 \quad KNO_2$$

6.16   Now that you know the answers, let's write the hydrolysis reactions for the three basic salts in Exercise 6.15.

_____

_____

_____

6.17   Consider now the hydrolysis of NaOAc.

a. Which ion is the spectator? _____

b. Which ion undergoes hydrolysis? _____

c. Which ion is hydrated? _____

d. Again, write out the hydrolysis reaction.

Acidic Salts. Just as there are salts of strong bases and weak acids, there are also salts of strong acids and weak bases. These are acidic salts. The most common examples, and the only ones we will use, contain the NH_4^+ cation—for instance, NH_4Cl. Although the system is a little more complex than implied here, we can imagine NH_4Cl to be formed by the reaction

$$NH_3(aq) + HCl(aq) \longrightarrow NH_4Cl(aq)$$

where $NH_3(aq)$ can be considered to be the weak base and HCl is the strong acid.

The arguments we have used with regard to basic salts apply to acidic salts as well, except that it is the cation rather than the anion that hydrolyzes. First we know that

$$NH_4Cl(s) + H_2O \longrightarrow \text{acidic solution}$$

The pH of such a solution will be less than 7 (acidic). This means that there is an excess of H_3O^+ ions present, and it is our goal to determine how these ions are generated.

Assuming, as usual, that salts are 100% ionized in aqueous solution, we obtain

$$NH_4Cl(s) + H_2O \longrightarrow NH_4^+(aq) + Cl^-(aq)$$

But the Cl^- ion is the anion of a strong acid. We saw in our discussion of neutral salts (p. 109) that such an ion merely undergoes hydration. On the other hand, the NH_4^+ ion is the cation of a weak base and will undergo hydrolysis. The acidity of the solution must therefore be attributed to the behavior of the NH_4^+ ion. The hydrolysis reaction is

$$NH_4^+(aq) + H_2O \rightleftharpoons NH_3(aq) + H_3O^+(aq)$$

In this case we see the NH_4^+ ion acting as a protonic acid and the H_2O molecule acting as a protonic base.

_____ **EXERCISE** _____

6.18 Let's determine what we can say about NH_4Br.

a. It is a salt and can be considered to be formed from the strong acid

_____ and the weak base _____ .

$$NH_3(aq) + HBr(aq) \longrightarrow \text{_____}$$

b. Salts ionize completely in aqueous solution. Thus the ionization of NH_4Br will give

$$NH_4Br(s) + H_2O \longrightarrow \text{_____} + \text{_____}$$

c. The Br^- ion undergoes hydration. Thus we have

$$Br^- + H_2O \longrightarrow \text{_____}$$

d. The NH_4^+ ion, being the cation of a _____ base, undergoes

_____ . The equation is

$$NH_4^+ + H_2O \rightleftharpoons \text{_____} + \text{_____}$$

e. NH_4Br is an _____ salt. Why? _____

_____

================================

As we did with the basic salts, it is also possible to write the complete ionic equation for the dissolution of an acidic salt.

$$NH_4^+(aq) + Cl^-(aq) + H_2O \rightleftharpoons NH_3(aq) + H_3O^+(aq) + Cl^-(aq)$$

Here the spectator ion is the Cl^- ion, and canceling it out, we arrive at the net ionic equation

$$NH_4^+(aq) + H_2O \longrightarrow NH_3(aq) + H_3O^+(aq)$$

which is the hydrolysis equation.

_____ **EXERCISES** _____

6.19 Circle the acidic salts among the following.

$$BaF_2 \quad NH_4NO_3 \quad Cd(ClO_3)_2 \quad RbCl \quad CuI_2$$

6.20 If you have really understood what is meant by the hydrolysis of a salt, the rest should be straightforward. Write all equations involved in the dissolution of the following salts in water. Tell if the solution is acidic, basic, or neutral, and list all spectator ions.

a. NaCN

b. NH_4NO_3

c. $RbClO_4$

7

Acid-Base III:
Lewis Acids and Bases

(Coordination Chemistry)

In Chapter 5, the Brønsted-Lowry (protonic) definition of acids and bases was introduced. There we saw that the Brønsted-Lowry definition limited acid-base behavior to species that accept or donate a proton (H^+ ion).

EXERCISES

7.1 As a review,

a. define a Brønsted-Lowry acid: _____

b. define a Brønsted-Lowry base: _____

7.2 In the reaction

$$HCl + H_2O \longrightarrow H_3O^+(aq) + Cl^-(aq)$$

a. the Brønsted-Lowry acid is _____ .

b. the Brønsted-Lowry base is _____ .

At about the same time that the protonic acid-base definition was first published, G. N. Lewis proposed a more fundamental and also much broader definition of acids and bases. According to Lewis, *an acid is any species capable of accepting an electron pair to form a covalent bond*, and a *base is any species capable of donating an electron pair to form a covalent bond.* The standard illustration of a Lewis acid-base reaction is

$$\begin{array}{ccc} \text{F} & \text{H} & \text{F} \quad \text{H} \\ | & | & | \quad | \\ \text{F}-\text{B} + :\text{N}-\text{H} & \longrightarrow & \text{F}-\text{B}-\text{N}-\text{H} \\ | & | & | \quad | \\ \text{F} & \text{H} & \text{F} \quad \text{H} \end{array}$$

--------- **EXERCISE** ---------

7.3 Note that BF_3 has a vacant orbital and may accept an electron pair.

Therefore, BF_3 is a Lewis _____ . NH_3, having an unused lone pair of electrons, can donate these to the vacant orbital on the boron. NH_3 is, therefore, a Lewis _____ .

The net result of a Lewis acid-base reaction is the formation of a covalent bond. The formation of this covalent bond, however, is a little different from the usual. Here both electrons in the electron pair are donated by one species, rather than one electron coming from each species. Such a bond is referred to as a *coordinate covalent bond.* Lewis acid-base reactions, then, involve the formation of a coordinate covalent bond. Once the bond is formed, it is the same as the usual type of covalent bond. It is only in terms of its origin that it is distinguished from a conventional covalent bond.

COMPARISON OF THE LEWIS AND THE BRØNSTED-LOWRY DEFINITIONS

In order to make a comparison between these two definitions, let's begin with the reaction

$$\text{H}^+ + :\overset{\text{H}}{\underset{..}{\text{O}}}-\text{H} \longrightarrow \left[\text{H}:\overset{\text{H}}{\underset{..}{\text{O}}}-\text{H} \right]^+$$

The Lewis definition, being so general, actually incorporates the Brønsted-Lowry definition. In the above reaction, water accepts the proton and is, therefore, a Brønsted-Lowry base. The important point is that the

H_2O is able to accept the proton because it has an available lone pair of electrons that it can donate to the *vacant orbital* on the H^+ ion. Thus, the H_2O molecule is also a Lewis base. The H^+ ion, having an available orbital, accepts the electron pair and is, therefore, a Lewis acid. But a technical difficulty arises here. It is the H^+ ion rather than the protonic acid HX that is the Lewis acid. Whereas HX is a protonic acid, it is not a Lewis acid. On the other hand, the H^+ ion is a Lewis acid, but it is not a protonic acid. However, recognizing that the source of the H^+ ion is the protonic acid, HX, most chemists ignore this distinction. If we overlook this point, it should be apparent that protonic acid-base reactions can be covered by the Lewis definition. The formation of any protonic acid from its anion and H^+ is, indeed, a Lewis acid-base reaction, as is the reaction of the H^+ ion with a molecule such as H_2O.

$$H^+(aq) + OAc^-(aq) \rightleftharpoons HOAc(aq)$$

$$H^+(aq) + H_2O \rightleftharpoons H_3O^+(aq)$$

Some Lewis Acid-Base Reactions. Some further examples of Lewis acid-base reactions are

Acid	Base	

$$H^+(aq) + F^-(aq) \rightleftharpoons HF(aq)$$

$$H^+(aq) + NH_3(aq) \rightleftharpoons NH_4^+(aq)$$

$$BF_3(aq) + F^-(aq) \rightleftharpoons BF_4^-(aq)$$

These examples illustrate the fact that Lewis acid-base chemistry incorporates both ionic and molecular species. But more important, note that the acid does *not* have to be the H^+ ion. Looking at the Lewis structures for the same reactions,

$$H^+ + :\overset{..}{\underset{..}{F}}:^- \rightleftharpoons H:\overset{..}{\underset{..}{F}}:$$

$$H^+ + :\overset{H}{\underset{H}{N}}:H \rightleftharpoons \left[H:\overset{H}{\underset{H}{N}}:H \right]^+$$

$$:\overset{:\overset{..}{F}:}{\underset{:\overset{..}{F}:}{F}}:B + :\overset{..}{\underset{..}{F}}:^- \longrightarrow \left[:\overset{:\overset{..}{F}:}{\underset{:\overset{..}{F}:}{F}}:B:\overset{..}{\underset{..}{F}}: \right]^-$$

we can see that, in each case, the Lewis base donates an electron pair to an electron deficient species, the Lewis acid.

—————— **EXERCISE** ——————

7.4 As a first try, give the products of the following reactions.

a. $H^+(aq) + OH^-(aq) \longrightarrow$ _____

b. $H^+(aq) + CN^-(aq) \longrightarrow$ _____

c. $HI(g) + PH_3(g) \longrightarrow$ _____

d. $BCl_3 + Cl^- \longrightarrow$ _____

Oxides as Lewis Acids or Bases. The same general type reactions can occur with oxides. For instance,

$$2\,NaOH(s) + SO_3(g) \longrightarrow Na_2SO_4(s) + H_2O$$

In this reaction a hydroxyl base reacts with an acid anhydride to form a salt and water. The products are identical to those obtained when NaOH reacts with the corresponding acid, H_2SO_4. The acid anhydride, SO_3, therefore is acting as an acid in this reaction. As it has no proton to donate, SO_3 cannot be a protonic acid, so it must be a Lewis acid. Note that SO_3 reacts to form $SO_4{}^{2-}$, thereby gaining an electron pair. The only species that possesses an available electron pair is the OH^- ion. Thus, the OH^- ion acts as the Lewis base. Using Lewis structures the reaction is

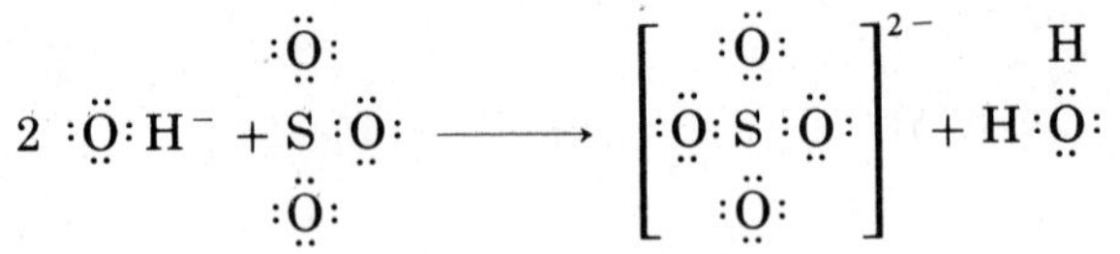

$$2\;:\!\ddot{O}\!:\!H^- + S\!:\!\ddot{O}\!: \longrightarrow \left[:\!\ddot{O}\!:\!S\!:\!\ddot{O}\!: \right]^{2-} + H\!:\!\ddot{O}\!:$$

Likewise we find

$$Na_2O(s) + H_2SO_4(l) \longrightarrow Na_2SO_4(s) + H_2O$$

where the basic anhydride reacts with the protonic acid to give a salt and water. Again, the same products would form if H_2SO_4 were reacting with the corresponding hydroxyl base, NaOH. The basic anhydride, then, is the base in this reaction. In this situation, the oxide ion is the Lewis base, as it is the only species that has an available electron pair. Thus,

$$:\ddot{O}:^{2-} + 2\ H^+ \longrightarrow H:\overset{\textstyle H}{\underset{\textstyle}{\ddot{O}}}:$$

Now, the same technical problem arises here as it did in the introductory discussion on Lewis acids. Most chemists, recognizing the source of the O^{2-} ion, would call Na_2O the Lewis base.

Finally, the oxides can react directly:

$$CaO(s) + CO_2(g) \longrightarrow CaCO_3(s)$$

The only difference in this case is the absence of H_2O as a product. Obviously if there is no hydrogen among the reactants, H_2O cannot possibly be formed. The generality of the Lewis definition is illustrated in this reaction: an acid and a base, neither containing hydrogen, react to form a salt. From the previous discussion in this section, it can be concluded that CO_2, the acid anhydride, is the Lewis acid, and O^{2-}, the only species capable of donating an electron pair, is the Lewis base.

_____ **EXERCISE** _____

7.5 Now, complete the following reactions assuming complete neutralization.

a. $HCl(g) + Na_2O(s) \longrightarrow$ _____

b. $BaO(s) + H_2SO_4(l) \longrightarrow$ _____

c. $CaO(s) + SO_3(g) \longrightarrow$ _____

d. $Ca(OH)_2(s) + CO_2(g) \longrightarrow$ _____

e. $MgO(s) + SO_2(g) \longrightarrow$ _____

f. $H_3PO_4(l) + CsO(s) \longrightarrow$ _____

g. $K_2O(s) + CO_2(g) \longrightarrow$ _____

h. $Sr(OH)_2(s) + CO_2(g) \longrightarrow$ _____

COMPLEX IONS

Now let's consider the reaction

$$Al^{3+} + 6 \ F^- \longrightarrow AlF_6{}^{3-}$$

At first glance, it may appear that there are not six available vacant orbitals on the Al^{3+} ion. Recall, however, that the electron configurations are

$$Al: \ [Ne] \ \underset{3s}{\overset{\uparrow\downarrow}{\rule{0.8em}{0.4pt}}} \ \underset{3p}{\overset{\uparrow}{\rule{0.8em}{0.4pt}} \ \rule{0.8em}{0.4pt} \ \rule{0.8em}{0.4pt}} \quad \text{and} \quad Al^{3+}: \ [Ne] \ \underset{3s}{\rule{0.8em}{0.4pt}} \quad \underset{3p}{\rule{0.8em}{0.4pt} \ \rule{0.8em}{0.4pt} \ \rule{0.8em}{0.4pt}}$$

Here we immediately see that Al^{3+} has four vacant orbitals. In addition, the five $3d$ orbitals of the ion are not being used. Two of these plus the s and the three p orbitals provide the necessary six orbitals (sp^3d^2) to accept the six electron pairs donated by the six F^- ions.

A species such as $AlF_6{}^{3-}$ is called a *complex ion*. Complex ions are defined as charged species in which a metal atom or metal ion is bonded to a group of neutral molecules and/or negative ions. We will limit this discussion to those complex ions in which the metal is bonded either to neutral molecules or to ions. That is, we will not consider complex ions in which both neutral molecules and negative ions are present together.

Transition metals are particularly well known for their tendency to form complex ions. The species that attaches to the central metal is called the *complexing agent* or *ligand*. The ligand will donate a lone pair of electrons, and thus will act as a Lewis base. Following are some examples of the formation of common complex ions:

$$Co^{3+}(aq) + 6 \ Cl^-(aq) \longrightarrow CoCl_6{}^{3-}(aq)$$

$$Ag^+(aq) + 2 \ NH_3(aq) \longrightarrow Ag(NH_3)_2{}^+(aq)$$

$$Cu^{2+}(aq) + 4 \ H_2O \longrightarrow Cu(H_2O)_4{}^{2+}(aq)$$

Note that in all of these reactions, the metal ion acts as the Lewis acid.

Complex ions are very important in the study of qualitative analysis. If a student neglects to learn the behavior of these ions, he or she may well be faced with a number of difficulties. For example, there is a particular reaction that often puzzles qualitative analysis students. AgCl is insoluble in water, so it is easily precipitated by adding HCl to a solution containing Ag^+:

$$Ag^+(aq) + Cl^-(aq) \longrightarrow AgCl(s)$$

Often the student adds an excess of HCl and observes the precipitate to disappear. AgCl is insoluble, and yet it dissolves! This behavior doesn't make any sense unless one is aware that a soluble complex ion is being formed:

$$AgCl(s) + Cl^-(aq) \xrightarrow{\text{H}_2\text{O}} AgCl_2^-(aq)$$

_____ **EXERCISE** _____

7.6 In the previous reaction, name

a. the Lewis acid _____

b. the Lewis base _____

In the complex ions we have considered so far, you might have detected a pattern developing. The Ag^+ ion is bound to two NH_3 molecules in $Ag(NH_3)_2^+$ and to two Cl^- ions in $AgCl_2^-$. Referring to the other examples we have seen, note that the Co^{3+} ion is attached to 6 Cl^- ions, the Al^{3+} ion is attached to 6 fluoride ions, and the Cu^{2+} ion is attached to 4 H_2O molecules. A good rule of thumb to remember (although some chemists do not like it) is that the number of attached ligands is two times the charge on the central ion. There are many exceptions to this rule, but it will be correct more times than not.

_____ **EXERCISE** _____

7.7 Now it's time to see if you have the idea. Complete the following reactions. Assume that an excess of the ligand is added in order to assure the formation of a complex ion. Also assume that the rule of thumb holds.

a. $Ag^+ + CN^- \longrightarrow$ _____

b. $Cr^{3+} + H_2O \longrightarrow$ _____

c. $Fe^{3+} + H_2O \longrightarrow$ _____

d. $Co^{3+} + NH_3 \longrightarrow$ _____

e. $Cu^{2+} + NH_3 \longrightarrow$ _____

f. $Pt^{2+} + Cl^- \longrightarrow$ _____

g. $Zn^{2+} + CN^-$ ⟶ _____

h. $Fe^{3+} + F^-$ ⟶ _____

i. $Ti^{3+} + H_2O$ ⟶ _____

j. $Cd^{2+} + I^-$ ⟶ _____

The formation of complex ions can be much more complicated than we have indicated here. There can be mixed ligands of negative ions and molecules such as

$$[Co(NH_3)_4Cl_2]^+$$

There are also many ligands that can bond to the metal ion through more than one site. That is, they can donate two or more electron pairs. Finally, there are many exceptions to our rule of thumb. Some examples are

$$Fe^{2+} + 6\ H_2O \longrightarrow Fe(H_2O)_6{}^{2+}$$

$$Au^{3+} + 4\ CN^- \longrightarrow Au(CN)_4{}^-$$

$$Mn^{2+} + 6\ H_2O \longrightarrow Mn(H_2O)_6{}^{2+}$$

$$Fe^{2+} + 6\ CN^- \longrightarrow Fe(CN)_6{}^{4-}$$

But in spite of all of this, you have seen and, we hope, learned how to predict the reactions for the formation of some of the more common examples of complex ions.

Nomenclature of Complex Ions. To begin with, we have seen that complex ions can be either cations or anions. When naming complex ions, the ligand is always named before the metal, and the name of the ligand is the same for both cationic and anionic complexes. On the other hand, the name of the metal differs in the two types of ions. Let's develop the nomenclature through example by first considering the anionic complex $CoCl_6{}^{3-}$. Negative ligands are given an -o ending; thus chloride becomes chloro. Since there are six chlorine atoms in $CoCl_6{}^{3-}$, they are referred to as *hexa*chloro. In a complex anion, the metal is usually given its English name and is always given an *-ate* ending followed by its oxidation state in Roman numerals enclosed in parentheses. And as mentioned earlier, the ligands are named first. Thus $CoCl_6{}^{3-}$ is called the hexachlorocobaltate(III) ion. Note that the name is written as one word. As further examples we have

$$CrF_6^{3-} \text{ is hexafluorochromate(III) ion.}$$
$$PtCl_6^{2-} \text{ is hexachloroplatinate(IV) ion.}$$

_____ **EXERCISES** _____

7.8 In the complex ion $Cr(CN)_6^{3-}$ there are six cyanide ions as ligands which are named _____. In the complex anion, Cr is called _____. Remembering that the oxidation state of the Cr must be included, the name of $Cr(CN)_6^{3-}$ is _____.

7.9 Now try some without help.

a. $PtCl_6^{2-}$ _____

b. $CoBr_6^{3-}$ _____

c. $Co(CN)_6^{3-}$ _____

As you already know from previous examples in this chapter, the number of ligands in a complex ion is not restricted to six. The prefixes used to denote the numbers of attached ligands are

_di_—2		_penta_—5	
tri—3		_hexa_—6	
tetra—4		_hepta_—7	

As you might expect, there are exceptions to the nomenclature rules. The metals known to the ancients are given their Latin names with the -_ate_ ending when such metals are present in anionic species. These metals are

Element	Symbol	Name
copper	Cu	cuprate
gold	Au	aurate
iron	Fe	ferrate
lead	Pb	plumbate
silver	Ag	argentate
tin	Sn	stannate

Some examples of complex ions with these metals are

$FeF_6{}^{3-}$	hexafluoroferrate(III) ion
$Ag(CN)_2{}^-$	dicyanoargentate(I) ion
$Au(CN)_2{}^-$	dicyanoaurate(I) ion

-------- **EXERCISE** --------

7.10   Here are some complex anions to practice on. Give the names for the following:

a. $AgCl_2{}^-$ _____

b. $Al(OH)_4{}^-$ _____

c. $Au(CN)_4{}^-$ _____

d. $Fe(CN)_6{}^{3-}$ _____

e. $Ni(CN)_4{}^{4-}$ _____

For neutral ligands, there is no distinctive ending. The two most common neutral ligands, H_2O and NH_3, are called aqua and ammine, respectively. Another relatively common simple ligand, CO, is called carbonyl.

In cationic complexes, the metal is always given its English name followed, as usual, by the oxidation state in parentheses, but the *-ate* ending is not used. Two such examples are

$Ni(NH_3)_6{}^{3+}$	hexamminenickel(III) ion
$Mn(H_2O)_6{}^{2+}$	hexaaquamanganese(II) ion

And further note the contrast to the naming of anionic complexes. $Cu(NH_3)_4{}^{2+}$ is tetraammine*copper*(II) ion. The English rather than the Latin name of the metal is used.

-------- **EXERCISE** --------

7.11   Now see if you can name these cationic complexes.

a. $Cu(H_2O)_4{}^{2+}$ _____

b. $Ag(NH_3)_2{}^+$ _____

c. $Cr(H_2O)_6{}^{3+}$ _____

d. $Fe(H_2O)_6{}^{2+}$ _____

e. $Ni(NH_3)_4{}^{2+}$ _____

Nomenclature of Coordination Compounds. You have now seen how to name a complex cation as well as a complex anion. The next thing to do is to put these together to name a coordination compound. Actually naming such a compound is extremely simple. One merely names the cation first and then the anion; for instance,

$K_4[Fe(CN)_6]$	potassium hexacyanoferrate(II)
$[Cr(NH_3)_6]Cl_3$	hexaamminechromium(III) chloride
$Na[AuCl_4]$	sodium tetrachloroaurate(III)
$[Ni(NH_3)_4](NO_3)_2$	tetramminenickel(II) nitrate

_____ EXERCISES _____

7.12 Here is your chance to show what you know. Give the names for these:

a. $Cr(H_2O)_6^{3+}$ _____

b. $Ni(CO)_4$ _____

c. $Zn(OH)_4^{2-}$ _____

d. $FeCl_6^{3-}$ _____

e. $[Ag(NH_3)_2]Cl$ _____

f. $[Ag(NH_3)_2]_2[PtCl_4]$ _____

g. $Na[Ag(CN)_2]$ _____

h. $Ni(CN)_4^{2-}$ _____

i. $CuBr_4^{2-}$ _____

j. $Ni(NH_3)_4^{2+}$ _____

7.13 If you need more practice, name these.

a. $Mn(H_2O)_6^{2+}$ _____

b. $Zn(CN)_4^{2-}$ _____

c. $Na_2[PtCl_6]$ _____

d. $[Co(NH_3)_6]Br_3$ _____

e. $[Ag(NH_3)_2]NO_3$ _____

f. CoF_6^{3-} _____

g. $Al[Fe(CN)_6]$ _____

h. $[Mn(H_2O)_6]Cl_2$ _____

i. $K_4[Ni(CN)_4]$ _____

j. $AlF_6{}^{3-}$ _____

REVIEW III

In this third and last review chapter, both acid-base definitions and the corresponding reactions, as presented in the last three chapters, will be discussed. Recall that in Chapter 5 you learned about acid-base behavior from the Brønsted-Lowry viewpoint. Next you saw that salts often exhibit acid-base behavior, and a rationale had to be developed. This was accomplished in Chapter 6 when the concept of hydrolysis was presented. You were then introduced to the general acid-base definition of G. N. Lewis, a definition that is not limited to species accepting or donating H^+ ions. Such species as complex ions and reactions of acidic and basic anhydrides can now be rationalized within the framework of acid-base chemistry.

Again, if you feel that you have mastered the material in these last three chapters, by all means go on to Chapter 8. It will most likely be worth your while, however, to at least look at portions of this review to make sure that you do understand acid-base chemistry.

COMPARISON OF THE LEWIS AND THE BRØNSTED-LOWRY DEFINITIONS

Two important acid-base definitions have been discussed at length. It is imperative that you be very familiar with both of these. The Brønsted-Lowry (protonic) definition usually lends itself quite well to the explanation of common and familiar acid-base reactions, so it is often discussed at length in freshman texts. The more general Lewis definition lends itself to the explanation of acid-base reactions involving ions and molecules that may or may not contain acidic H atoms.

————— **EXERCISES** —————

III.1 Define a

a. Brønsted-Lowry acid

———————————————————————————————

b. Brønsted-Lowry base

———————————————————————————————

c. Lewis acid

———————————————————————————————

d. Lewis base

———————————————————————————————

III.2 Given the equation

$$HI(aq) + H_2O \longrightarrow H_3O^+(aq) + I^-(aq)$$

identify the

a. Brønsted-Lowry acid ———————————

b. Brønsted-Lowry base ———————————

c. Lewis acid ———————————

d. Lewis base ———————————

III.3 In the equation

$$BF_3 + F^- \longrightarrow BF_4^-$$

where there are no H^+ ions, and thus no Brønsted-Lowry acid or base, identify the

a. Lewis acid ———————————   **b.** Lewis base ———————————

================================

PREPARATION OF BRØNSTED-LOWRY ACIDS

The Brønsted-Lowry acids can be prepared by a number of methods. The simplest of these methods to remember are the direct combination reactions. These reactions include the preparation of binary acids by reacting H_2 with a nonmetal and the preparation of oxyacids from water and an acid anhydride.

_____ **EXERCISE** _____

III.4 Let's see what you remember. Give the acid products of the following.

a. $H_2(g) + F_2(g) \longrightarrow$ _____   **b.** $H_2(g) + Br_2(g) \longrightarrow$ _____

c. $H_2(g) + I_2(g) \longrightarrow$ _____   **d.** $H_2(g) + Cl_2(g) \longrightarrow$ _____

e. $CO_2(g) + H_2O \longrightarrow$ _____   **f.** $SO_3(g) + H_2O \longrightarrow$ _____

g. $SeO_2(s) + H_2O \longrightarrow$ _____   **h.** $I_2O_5(s) + H_2O \longrightarrow$ _____

========

Brønsted-Lowry (or protonic) acids can also be prepared by metathetical reactions. One common method involves the reaction of a salt with a less volatile acid (usually H_2SO_4). In a second method, a phosphorous halide or oxyhalide reacts with water to produce the desired acid.

_____ **EXERCISE** _____

III.5 Give both the acids and the by-products that are formed in the following reactions.

a. $KNO_3(s) + H_2SO_4(l) \xrightarrow{\Delta}$ _____

b. $NaBr(s) + H_2SO_4(l) \xrightarrow{\Delta}$ _____

c. $KClO_4(s) + H_2SO_4(l) \xrightarrow{\Delta}$ _____

d. $KCl(s) + H_2SO_4(l) \xrightarrow{\Delta}$ _____

e. $Ca(ClO_3)_2(s) + H_2SO_4(l) \xrightarrow{\Delta}$ _____

f. $PBr_3(l) + H_2O \longrightarrow$ _____

g. $PI_3(s) + H_2O \longrightarrow$ _____

h. $PCl_3(l) + H_2O \longrightarrow$ _____

i. $POCl_3(l) + H_2O \longrightarrow$ _____

j. $POBr_3(l) + H_2O \longrightarrow$ _____

========

BASIC ANHYDRIDES AND WATER

Bases can be prepared by reacting basic anhydrides with water. These reactions may be reversed upon heating.

────────── **EXERCISE** ──────────

III.6   Give the products of the following reactions:

a. $K_2O(s) + H_2O \longrightarrow$ _____

b. $SrO(s) + H_2O \longrightarrow$ _____

c. $Rb_2O(s) + H_2O \longrightarrow$ _____

d. $BaO(s) + H_2O \longrightarrow$ _____

e. $Al(OH)_3(s) \xrightarrow{\Delta}$ _____

f. $Fe(OH)_3(s) \xrightarrow{\Delta}$ _____

g. $Sr(OH)_2(s) \xrightarrow{\Delta}$ _____

h. $KOH(s) \xrightarrow{\Delta}$ _____

══════════════════════════════

NEUTRALIZATION

A solution is considered to be neutral when the concentrations of H_3O^+ ions and OH^- ions are equal. This occurs in water where the slight dissociation produces an equal amount of these ions. Neutralization also takes place when equal concentrations of strong protonic acids and bases are added together. In polyprotic acids, remember, all the acid hydrogens will be replaced if ample base is present. Similarly, all of the OH^- groups in a polyhydroxyl base will likewise be replaced if sufficient acid is present.

────────── **EXERCISE** ──────────

III.7   Give the products of these neutralization reactions. Assume complete replacement for any reactions involving a polyprotic or polyhydroxyl species.

a. $NaOH(aq) + HCl(aq) \longrightarrow$ _____

b. $KOH(aq) + HBr(aq) \longrightarrow$ _____

c. $CsOH(aq) + HI(aq) \longrightarrow$ _____

d. $RbOH(aq) + H_2SO_4(aq) \longrightarrow$ _____

e. $Ba(OH)_2(aq) + HCl(aq) \longrightarrow$ _____

f. $Sr(OH)_2(aq) + H_2SO_4(aq) \longrightarrow$ _____

g. $H_3PO_4(aq) + KOH(aq) \longrightarrow$ _____

h. $H_3PO_4(aq) + Ba(OH)_2(aq) \longrightarrow$ _____

BRØNSTED-LOWRY ACID-BASE REACTIONS

In reactions involving the dissociation of weak acids in water, a state of equilibrium will occur. Thus, one must recognize that both forward and reverse reactions exhibit acid-base behavior. Using the dissociation of HF as an example,

$$HF(aq) + H_2O \rightleftharpoons H_3O^+(aq) + F^-(aq)$$

HF is the acid and H_2O is the base. In the reverse reaction, H_3O^+ is the acid and F^- is the base. Thus two acid-base pairs need to be considered. To distinguish these pairs, the acid and base products (or the acid and base of the reverse reaction) are called a conjugate acid-base pair, whereas the original reactants are called simply the acid-base pair.

Often, a given substance may act as either an acid or a base. Water, for example, is acidic in the presence of NH_3, whereas it is basic in the presence of HCl.

It is a relatively simple matter to identify a Brønsted-Lowry acid. After all, it must have a proton! The identification of bases, however, is somewhat more difficult. We know, of course, that Brønsted-Lowry bases are proton acceptors, but the problem lies in the identification of the species that actually accepts the proton. Three rules that are helpful for remembering the basic species in a protonic acid-base reaction are

1. Anions of weak acids are bases in the presence of water.
2. Ammonia is a base in the presence of water.
3. Water is a base in the presence of substances that are commonly called acids.

_____ **EXERCISE** _____

III.8 Now, complete the following reactions and identify the acid and the conjugate base.

	Acid	Conjugate base
a. $HCN(aq) + H_2O \rightleftharpoons$	_____	_____
b. $HClO(aq) + H_2O \rightleftharpoons$	_____	_____
c. $HF(aq) + H_2O \rightleftharpoons$	_____	_____
d. $H_2SO_3(aq) + H_2O \rightleftharpoons$	_____	_____
e. $H_3PO_4(aq) + H_2O \rightleftharpoons$	_____	_____
f. $CN^-(aq) + H_2O \rightleftharpoons$	_____	_____
g. $NO_2^-(aq) + H_2O \rightleftharpoons$	_____	_____

SALTS

There are three types of salts: neutral, acidic, and basic. Neutral salts are formed from strong acids and strong bases. Basic salts are formed from strong bases and weak acids. Acidic salts are formed from strong acids and weak bases. Neutral salts are neutral in solution because no net increase of either H_3O^+ or OH^- ions occurs when the salt dissolves. The ions of the neutral salt merely undergo *hydration*. One might say that no new species are formed in the solution. New species are formed, however, when acidic or basic salts dissolve in water. In the case of a basic salt, its anion reacts with water to form OH^- ions, whereas the cation of the acidic salt reacts with water to form H_3O^+ ions. The process in which either H_3O^+ or OH^- ions are formed by the splitting of a water molecule is called *hydrolysis*.

EXERCISE

III.9   Give the products of these hydrolysis reactions.

a. $F^-(aq) + H_2O \rightleftharpoons$ _____

b. $CN^-(aq) + H_2O \rightleftharpoons$ _____

c. $NO_2^-(aq) + H_2O \rightleftharpoons$ _____

d. $NH_4^+(aq) + H_2O \rightleftharpoons$ _____

e. $HCO_3^-(aq) + H_2O \rightleftharpoons$ _____

f. $HPO_4^{2-}(aq) + H_2O \rightleftharpoons$ _____

g. $HSO_4^-(aq) + H_2O \rightleftharpoons$ _____

h. $OAc^-(aq) + H_2O \rightleftharpoons$ _____

LEWIS ACID-BASE REACTIONS

The Lewis definition, which treats acids as electron pair acceptors and bases as electron pair donors, is more general than the Brønsted-Lowry definition. Acid behavior is no longer limited to proton-containing compounds. A Lewis acid, then, may be a proton, metal ion, acid anhydride, or any other electron-deficient species. A base may be any species possessing an available electron pair.

EXERCISE

III.10   Identify the Lewis acid and Lewis base in each of the following reactions.

	Lewis acid	Lewis base
a. $H^+(aq) + OH^-(aq) \longrightarrow H_2O$	_____	_____
b. $NH_3(aq) + H^+(aq) \longrightarrow NH_4^+(aq)$	_____	_____
c. $BF_3(aq) + F^-(aq) \longrightarrow BF_4^-(aq)$	_____	_____
d. $BF_3(g) + NH_3(g) \longrightarrow BF_3NH_3(s)$	_____	_____
e. $SbF_5(aq) + F^-(aq) \longrightarrow SbF_6^-(aq)$	_____	_____
f. $BaO(s) + CO_2(g) \longrightarrow BaCO_3(s)$	_____	_____
g. $Cu^{2+}(aq) + 4\,NH_3(aq) \longrightarrow Cu(NH_3)_4^{2+}(aq)$	_____	_____
h. $Zn^{2+}(aq) + 4\,OH^-(aq) \longrightarrow Zn(OH)_4^{2-}(aq)$	_____	_____

Transition metal ions readily form complex ions. This tendency on the part of these ions is so pronounced that a branch of chemistry called coordination chemistry exists.

---------   **E X E R C I S E**   ---------

III.11   Give the complex ions formed in the following reactions. Remember to employ the rule of thumb when determining the number of ligands attached to the metal ion.

a. $Cr^{3+}(aq) + H_2O \longrightarrow$ _____

b. $Cu^{2+}(aq) + NH_3(aq) \longrightarrow$ _____

c. $Ag^+(aq) + CN^-(aq) \longrightarrow$ _____

d. $Fe^{3+}(aq) + F^-(aq) \longrightarrow$ _____

e. $Cr^{3+}(aq) + F^-(aq) \longrightarrow$ _____

f. $Co^{2+}(aq) + SCN^-(aq) \longrightarrow$ _____

g. $Pt^{2+}(aq) + Cl^-(aq) \longrightarrow$ _____

NOMENCLATURE OF BRØNSTED-LOWRY ACIDS

There are two kinds of acids with which we are concerned: hydroacids and oxyacids. Hydroacids are named by adding the prefix *hydro-* and the suffix *-ic* to the stem of the non-hydrogen element along with the word acid. Thus, HCl is called *hydrochloric* acid. Oxyacid nomenclature, on the other hand, depends upon the oxidation state of the central atom. In the simple cases, those in which the central atom has only two oxidation states, the rules are as follows. For the acid whose central atom is in its higher oxidation state, merely add an *-ic* ending to the stem of the central atom along with the word acid. If the central atom is in its lower oxidation state, repeat the previous procedure with the ending *-ous*. Thus, HNO_3 and HNO_2 are nit*ric* and nit*rous* acid, respectively. In the case of the oxyacids of the halogens, in which the central atom possesses four oxidation states, assign the *-ic* and *-ous* endings to the two acids with the middle oxidation states. The acid whose central atom is in the highest oxidation state is assigned a *per-* prefix and an *-ic* ending, whereas the acid with the lowest oxidation state is given a *hypo-* prefix and an *-ous* ending. Thus we have

$\overset{7+}{H Br O_4}$	perbromic acid
$\overset{5+}{H Br O_3}$	bromic acid
$\overset{3+}{H Br O_2}$	bromous acid
$\overset{1+}{H Br O}$	hypobromous acid

E X E R C I S E

III.12   Now, name the following acids:

a. HBr _____

b. H_2S _____

c. H_3PO_3 _____

d. HClO _____

e. H_2CrO_4 _____

f. H_3AsO_4 _____

g. H_2SeO_3 _____

h. $HAlO_2$ _____

i. HI _____

j. $HClO_3$ _____

and the exception:

k. HCN _____

NOMENCLATURE OF COMPLEX IONS

When naming complex ions, remember that the ligands are always named first, preceded by the prefix denoting the number of ligands in the ion. Also, remember that English names are given to the metal ion[1] and that the charge on that ion is denoted by Roman numerals in parentheses immediately following the name of the ion. The only difference in the naming of cations and anions is that for anions an *-ate* ending is added to the stem of the central ion. Charged ligands are given an *-o* ending, whereas neutral ligands have distinctive names. For example, CO is called carbonyl, H_2O is called aqua, and NH_3 is called ammine.

E X E R C I S E

III.13   As the last exercise in this review, name the following ions.

a. $CuCl_4^{2-}$ _____

b. $Fe(CN)_6^{3-}$ _____

c. $Zn(CN)_4^{2-}$ _____

d. CdI_4^{2-} _____

e. $AgCl_2^-$ _____

f. $AuCl_4^-$ _____

g. $Co(NH_3)_6^{3+}$ _____

h. $Fe(H_2O)_6^{2+}$ _____

i. $Ag(NH_3)_2^+$ _____

j. $Ni(NH_3)_4^{2+}$ _____

[1] Remember that for anions the metals known to the ancients are given Latin names. If you've forgotten those names, then see p. 127.

8

Redox Reactions

A reaction that seems to fascinate new chemistry students is one in which a strip of copper metal is immersed in a colorless $AgNO_3$ solution. Silver metal deposits on the copper strip, and the solution turns blue. The reaction can be written as follows:

$$Cu(s) + 2\ AgNO_3(aq) \longrightarrow 2\ Ag(s) + Cu(NO_3)_2(aq)$$

In net ionic form, the equation is

$$Cu(s) + 2\ Ag^+(aq) \longrightarrow 2\ Ag(s) + Cu^{2+}(aq)$$

Note the changes in oxidation states:

$$Cu^0 \longrightarrow Cu^{2+} \quad \text{and} \quad Ag^+ \longrightarrow Ag^0$$

Oxidation takes place when a species undergoes an *increase* in oxidation state, and *reduction* takes place whenever a species undergoes a *decrease* in oxidation state. Thus, in this reaction, the Cu is *oxidized* and the Ag^+ ion is *reduced*. When such changes in oxidation states occur in a reaction, the reaction is called a *redox* reaction, which is short for an oxidation-reduction reaction.

For every oxidation there must be a corresponding reduction. This results from the fact that redox reactions involve the transfer of electrons from one species to another. Thus, for one species to gain electrons, some other species must lose them. The number of electrons gained, of course, must equal the number lost. In the previous reaction, we have

$$Cu^0 \longrightarrow Cu^{2+} + 2\,e^-$$

and
$$2\,e^- + 2\,Ag^+ \longrightarrow 2\,Ag^0$$

which shows that each Cu atom loses 2 electrons and each Ag^+ ion gains 1 electron. It is interesting to note that the species that is oxidized *loses* electrons while the species that is reduced *gains* electrons.

_____ **EXERCISE** _____

8.1 For an easy question, identify the species in the previous reaction that

a. is oxidized _____

b. is reduced _____

c. gains electrons _____

d. loses electrons _____

==

There is another important consideration here. The Ag^+ ion will not reduce by itself. It needs another species to donate the electrons to it. The metallic Cu supplies those electrons and is therefore called the *reducing agent*. Similarly, Cu does not oxidize by itself. It needs to donate its electrons to something. In this case, the electrons are donated to the Ag^+ ion, and, as might be expected, the Ag^+ ion is called the *oxidizing agent*. Note that the oxidized species is always the reducing agent and the reduced species is always the oxidizing agent.

It might appear that we are introducing a totally new type of reaction here, but that is not the case. In discussing direct combination reactions in Chapters 1 and 2, we were, in fact, dealing with the simplest type of redox reaction. This becomes apparent in the next exercise.

_____ **EXERCISES** _____

8.2 Some practice: consider the reaction

$$2\,\overset{0}{Na} + \overset{0}{Cl_2} \longrightarrow 2\,\overset{1+\;1-}{NaCl}$$

and identify the

a. oxidized species _____

b. reduced species _____

c. oxidizing agent _____

d. reducing agent _____

e. species that gains electrons _____

f. species that loses electrons _____

8.3   Now let's see what we've learned! In the following reactions identify (I) the species oxidized, (II) the species reduced, (III) the oxidizing agent, and (IV) the species that gains electrons.

	(I)	(II)	(III)	(IV)
a. $Mg + Cl_2 \longrightarrow MgCl_2$	_____	_____	_____	_____
b. $4\ Cr + 3\ O_2 \longrightarrow 2\ Cr_2O_3$	_____	_____	_____	_____
c. $Fe + 2\ HCl \longrightarrow FeCl_2 + H_2$	_____	_____	_____	_____
d. $2\ Na + ZnCl_2 \longrightarrow$ $2\ NaCl + Zn$	_____	_____	_____	_____
e. $8\ HNO_3 + 3\ Cu \longrightarrow$ $3\ Cu(NO_3)_2 + 2\ NO + 4\ H_2O$	_____	_____	_____	_____

OXIDIZING AND REDUCING AGENTS

Predicting whether a given species will act as an oxidizing or as a reducing agent requires a knowledge of the possible oxidation states of that species. For example, potassium exists either as the free element where it has an oxidation state of 0 or in a compound where it has an oxidation state of $1+$. No other oxidation state would be expected. Thus, elemental potassium has only one possibility: it must go to the $1+$ state and, therefore, act as a reducing agent, as can be seen in the following example:

$$\overset{0}{2\ K} + \overset{2+}{MgCl_2} \longrightarrow \overset{0}{Mg} + \overset{1+}{2\ KCl}$$

Here we see that K is oxidized and is, therefore, a reducing agent—in this case, reducing $\overset{2+}{Mg}$ to $\overset{0}{Mg}$.

Elemental potassium can never act as an oxidizing agent. To do so would require that the potassium be reduced, and there are no negative oxidation states of potassium. The Cl^- ion is also in its lowest oxidation state. Therefore Cl^- cannot go to a lower oxidation state, and it too can only be oxidized. Consequently, Cl^- can only act as a reducing agent. It turns out to be a very poor reducing agent, but, nevertheless, that is its only possibility. The following is an example of the Cl^- ion acting as a reducing agent:

$$2 \overset{1-}{Na}Cl + \overset{0}{F_2} \longrightarrow 2 Na \overset{1-}{F} + \overset{0}{Cl_2}$$

In this reaction, the Cl^- ion is the oxidized species and F_2 is the oxidizing agent. Correspondingly, F_2 is reduced, and the Cl^- ion is the reducing agent.

The issue of whether an element will act as an oxidizing agent or as a reducing agent becomes a little more complicated when the element can exist in several different oxidation states. Sulfur has the predicted oxidation states of $6+$, $4+$, 0, and $2-$. Elemental sulfur may, therefore, proceed to either a higher or a lower oxidation state:

$$\overset{0}{S} \longrightarrow \overset{6+}{S} \quad or \quad \overset{4+}{S} \quad and \quad \overset{0}{S} \longrightarrow \overset{2-}{S}$$

thereby acting as either an oxidizing or a reducing agent. Which of these possibilities will actually occur depends on the reaction, as can be seen from the following examples:

$$\overset{0}{S_8} + 8 O_2 \longrightarrow 8 \overset{4+}{S}O_2$$
$$16 Na + \overset{0}{S_8} \longrightarrow 8 Na_2 \overset{2-}{S}$$

In the first reaction sulfur is acting as a reducing agent, whereas in the second it is acting as an oxidizing agent.

In order to know the redox behavior of a particular species, one must first know the possible oxidation states that it may have.

EXERCISES

8.4   See if you've gotten the idea.

	Higher oxidation states	Lower oxidation states	Type of agent
a. Cr^{3+}	6+	0	both
b. Na^+	none	0	oxidizing
c. S^{2-}			
d. Se			
e. Cl^{7+}			
f. Ca^{2+}			
g. O_2^{2-}			
h. O^{2-}			

8.5 Try these for something a little different.

a. Circle those ions that can act *only* as reducing agents.

$$Mg^{2+} \quad N^{3-} \quad Cd^0 \quad I^- \quad O^{2-} \quad La^{3+}$$

b. Circle those which can act *only* as oxidizing agents.

$$Al^{3+} \quad Mn^{7+} \quad Mn^{3+} \quad Ce^0 \quad S^{2-} \quad F_2^0$$

ACTIVITY SERIES OF THE METALS

Let's now return to the reaction we discussed at the beginning of this chapter,

$$Cu^0(s) + 2\,AgNO_3(aq) \longrightarrow Cu(NO_3)_2(aq) + 2\,Ag^0(s)$$

First we need to answer the question, "Is this reaction reversible?" That is, will the following reaction take place?

$$2\,Ag^0(s) + Cu(NO_3)_2(aq) \longrightarrow Cu^0(s) + 2\,AgNO_3(aq)$$

Experimentally, it is found that when a strip of silver metal is placed in a solution of $Cu(NO_3)_2$, no reaction occurs; i.e.,

$$2\,Ag^0(s) + Cu(NO_3)_2(aq) \longrightarrow \text{no reaction}$$

Apparently, Cu^0 reduces Ag^+, but Ag^0 does not reduce Cu^{2+}. This *one-way* behavior is the general pattern for redox reactions. Here is another example:

$$\overset{0}{Zn}(s) + 2\,\overset{1+}{H}Cl(aq) \longrightarrow \overset{2+}{Zn}Cl_2(aq) + \overset{0}{H_2}(g)$$

When zinc is placed in an HCl solution, hydrogen gas is evolved, but the reverse reaction does not occur:

$$ZnCl_2(aq) + H_2(g) \longrightarrow \text{no reaction}$$

Again, note the pattern. Zinc reduces H^+, but hydrogen does not reduce Zn^{2+}. Redox reactions tend to go in one direction. It would be nearly impossible to memorize which one of a metal–metal ion pair in a reaction mixture tends to be the oxidizing agent and which tends to be the reducing agent. These roles, however, can be determined by making use of the *activity series* of the metals. With proper use of this series you will be able to predict which free metals can reduce a given metallic ion in a compound.

The Activity Series.   The activity series is a listing of metals in the order of their relative reducing abilities. An abbreviated form of the series is shown in Table 8.1. Note in particular that

1. The alkali and alkaline earth metals (families IA and IIA) are at the top of the list. These elements have very low ionization energies, and although this is not the only factor involved, it is extremely important. They readily lose electrons and are, therefore, good reducing agents.
2. Hydrogen serves to divide the series. Those elements above H_2 in the series, being better reducing agents than H_2, will reduce the H^+ ion, whereas those below hydrogen do not reduce the H^+ ion. This will prove to be important when we consider the reactions of metals with acids.
3. The elements below hydrogen are very inert metals. Their lack of reactivity is evident from the fact that most of these are used to make jewelry.[1]
4. Since the elements are listed in terms of their relative reducing abilities, any element is capable of reducing the cation in a compound if the cation is below it in the series. Thus, for instance, Li metal can reduce the K^+ ion in a compound such as KCl.

---

[1] Cu, Ag, and Au are often referred to as the *coinage metals* because of their historic use in making coins.

Table 8.1 Activity Series of the Metals

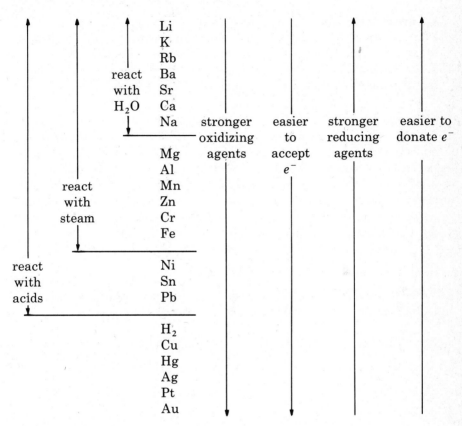

Reactions with Acids. By means of the activity series, one can predict whether or not a metal can react with an acid by reducing the H^+ ion. If a metal is above H_2 in the activity series, it is a better reducing agent than H_2 and should reduce the H^+ ion. Thus,

$$\overset{0}{Fe}(s) + \overset{1+}{HCl}(aq) \longrightarrow \overset{2+}{FeCl_2}(aq) + \overset{0}{H_2}(g)$$

whereas

$$Ag(s) + HCl(aq) \longrightarrow \text{no reaction}$$

To analyze these reactions note that

1. We are considering the reaction of a metal with an acid.
2. Iron is above H_2 in the activity series and, consequently, is a better reducing agent than H_2. This means that Fe can reduce the H^+ ion.
3. Silver is below H_2 in the activity series and, consequently, is a poorer reducing agent than H_2. This means that Ag cannot reduce the H^+ ion.

Although there are some borderline cases, almost all of the metals are above H_2 in the activity series. Those below H_2 are generally metals used to make jewelry and most are well known to the student. Thus, in most instances, the result of a redox reaction between a metal and an acid should be easy to predict, even without the activity series. In fact, in the next exercise you will have an opportunity to see how well you understand these principles. And, although you will be permitted to use the activity series, first see if you can give the results without looking at it.

_____ **EXERCISE** _____

8.6 Use the activity series, if necessary, to complete the following aqueous reactions. If no reaction occurs, so indicate with NR.

a. $Na + HCN \longrightarrow \underline{NaCN + H_2}$ **b.** $Cr + HI \longrightarrow$ _____

c. $Au + HBr \longrightarrow$ _____   **d.** $Pt + HI \longrightarrow$ _____

e. $Zn + HNO_2 \longrightarrow$ _____   **f.** $K + HF \longrightarrow$ _____

g. $Ag + H_3PO_4 \longrightarrow$ _____   **h.** $Sn + HCl \longrightarrow$ _____

i. $Cd + H_2SO_3 \longrightarrow$ _____   **j.** $Mg + H_3PO_4 \longrightarrow$ _____

Reactions with Water. In our discussion of acids and bases, we saw that H_2O can act as an acid. It is quite reasonable, then, to expect that some metals can reduce the H^+ ion present in H_2O, just as they do with HCl. It turns out that only the most reactive of the metals will behave in this manner with cold water, and a few more will do so with steam. It is necessary to have a copy of the activity series in order to know for certain which metals will reduce the H^+ ion in H_2O, but one can readily remember that the elements in families **IA** and **IIA** will do so. Note the similarities between the two reactions below:

$$\overset{0}{Na} + \overset{1+}{H}Cl \longrightarrow \overset{1+}{Na}Cl + \overset{0}{H_2}$$

$$\overset{0}{Na} + \overset{1+}{H}(OH) \longrightarrow \overset{1+}{Na}(OH) + \overset{0}{H_2}$$

_____ **EXERCISE** _____

8.7 These should be easy, but use the activity series to be sure. Complete the equations; $H_2O(l)$ means liquid H_2O and $H_2O(g)$ means steam. If no reaction occurs, so indicate with NR.

a. $Ca + H_2O(l) \longrightarrow$ _____   **b.** $Zn + H_2O(g) \longrightarrow$ _____

c. $Pt + H_2O(g) \longrightarrow$ _____   **d.** $Mn + H_2O(l) \longrightarrow$ _____

e. $Cr + H_2O(g) \longrightarrow$ _____   **f.** $Ag + H_2O(g) \longrightarrow$ _____

g. $Sr + H_2O(l) \longrightarrow$ _____   **h.** $Sn + H_2O(l) \longrightarrow$ _____

i. $K + H_2O(l) \longrightarrow$ _____   **j.** $Hg + H_2O(g) \longrightarrow$ _____

_____

Reactions with Metal Cations.   Now let's extend our use of the activity series beyond the reactions of metals with the H^+ ion. First, consider the reaction

$$2\,\overset{0}{Li}(s) + \overset{2+}{Zn}Cl_2(aq) \longrightarrow 2\,\overset{1+}{Li}Cl(aq) + \overset{0}{Zn}(s)$$

In this reaction, Li reduces the Zn^{2+} ion, and, inasmuch as Li is above Zn in the activity series, we should expect the reaction to take place as written. It might be easier to visualize if we write the reaction as

$$2\,Li^0(s) + Zn^{2+}(aq) \longrightarrow 2\,Li^+(aq) + Zn^0(s)$$

But we would *not* expect

$$Zn^0(s) + 2\,Li^+(aq) \longrightarrow 2\,Li^0(s) + Zn^{2+}(aq)$$

and, in fact, the latter reaction does not take place. We can also look at the reaction in terms of the donation and acceptance of electrons:

$$2 \, \text{Li}(s) \longrightarrow 2 \, \text{Li}^+(aq) + 2 \, e^- \qquad \text{and} \qquad 2 \, e^- + \text{Zn}^{2+}(aq) \longrightarrow \text{Zn}(s)$$

Now consider the reaction between Ca and $BaCl_2$. We can write it down as

$$\text{Ca}(s) + \text{BaCl}_2(aq) \longrightarrow \text{Ba}(s) + \text{CaCl}_2(aq)$$

But does this reaction really take place? Writing it down doesn't make it so, as many rueful test-takers well know! The activity series tells us that barium is the better reducing agent. Consequently, we must conclude that the reaction does *not* take place as written. Rather, we should write

$$\text{Ca}(s) + \text{BaCl}_2(aq) \longrightarrow \text{NR}$$

Knowing that barium is above calcium in the activity series, by now you should have figured out that the reaction that actually occurs is

$$\text{Ba}(s) + \text{CaCl}_2(aq) \longrightarrow \text{Ca}(s) + \text{BaCl}_2(aq)$$

———————— **EXERCISE** ————————

8.8   It is time to review again. With respect to the above reaction between Ba and $CaCl_2$, circle the correct response.

a. Since Ba is the (oxidizing / ⟨reducing⟩) agent, it (gains / loses) electrons.
b. Inasmuch as Ca^{2+} is the (oxidizing / reducing) agent, it is (oxidized / reduced).
c. Ba is above Ca in the activity series; therefore, Ba is a better (oxidizing / reducing) agent than Ca.
d. Since Ba (gains / loses) electrons, it (increases / decreases) in positive oxidation state.

Generalizations about the Activity Series.   We can now make some generalizations about the activity series. These generalizations will enable us to make good guesses concerning the outcome of a reaction without actually looking at the activity series itself. To accomplish this, it is necessary to divide the series into three groups as follows:

Group 1. This group includes the alkali and the alkaline earth metals (families **IA** and **IIA**) which are the best reducing agents. This grouping will be at the top of the series.

Group 3. The so-called jewelry metals (Cu, Ag, Au, and Pt) are included in this grouping. Mercury, being a liquid, is obviously not a jewelry metal, but it also falls in this grouping. These are relatively inert elements as is indicated by their tendency to remain in the free metallic state. They are the poorest reducing agents and are at the bottom of the activity series.

Group 2. This grouping includes all of the other metals. They are poorer reducing agents than those listed in Group 1 but better reducing agents than the jewelry metals listed in Group 3.

1	Li, K, Rb, Ba, Sr, Ca, Na, Mg
2	Al, Mn, Zn, Cr, Fe, Ni, Sn, Pb
	H₂
3	Cu, Hg, Ag, Pt, Au

   Metals in Group 1 can reduce any metal ions in Groups 2 and 3. And, as you should by now expect, metals in Group 2 can only reduce those metal ions in Group 3. Finally, metals in Group 3 are incapable of reducing metal ions from either Group 1 or Group 2. To illustrate each of these cases, let's look at the following aqueous reactions.

$$2 \, Na + ZnCl_2 \longrightarrow Zn + 2 \, NaCl$$

$$Na + AgCN \longrightarrow Ag + NaCN$$

$$Cr + KCl \longrightarrow NR$$

$$Cr + Au(CN)_3 \longrightarrow Au + Cr(CN)_3$$

$$Ag + KCl \longrightarrow NR$$

$$Ag + FeF_2 \longrightarrow NR$$

Based on the above groupings, we know that Na (Group 1) is above Zn (Group 2) and Ag (Group 3) and will, therefore, reduce their ions. Next we note that Cr is in Group 2, whereas K is in Group 1. Thus, Cr will not reduce the K^+ ion, and we observe no reaction between Cr and KCl. On the other hand, Cr is above Au (Group 3), and here we observe that Cr does reduce the Au^{3+} ion. Finally, Ag is in Group 3, and it cannot reduce the K^+ ion (Group 1) or the Fe^{2+} ion (Group 2). Consequently, no reaction takes place in either of these last two cases. Of course, when both metals in a proposed reaction come from the same grouping, you will probably find it necessary to look at the activity series.

Two further generalizations can be made. All of the Group 1 metals, except Mg, react with $H_2O(l)$ by reduction of the H^+ ion, and all of the metals in Groups 1 and 2 can react with acids by reducing the H^+ ion. With a knowledge of these generalizations, we can determine the fates of a great many reactions without looking at the activity series. For example,

$$Ag(s) + H_2O(l) \longrightarrow NR$$

Since Ag is in Group 3, it does not react with $H_2O(l)$ by reduction of the H^+ ion. Of course, you surely know this. Otherwise, you would never wash your hands while wearing a silver ring. On the other hand,

$$Ca(s) + 2\ H_2O \longrightarrow Ca(OH)_2(aq) + H_2(g)$$

Calcium is a family IIA metal (Group 1), and we now know that it does react with $H_2O(l)$ by reduction of the H^+ ion.

Next, let's consider the reaction of a member from each of the groupings with an acid. From our generalization, we would expect and, in fact, do observe the following:

Group 1: $2\ Na(s) + 2\ HCl(aq) \longrightarrow 2\ NaCl(aq) + H_2(g)$

Group 2: $Fe(s) + 2\ HCl(aq) \longrightarrow FeCl_2(aq) + H_2(g)$

Group 3: $Ag(s) + HCl(aq) \longrightarrow NR$

EXERCISES

8.9 Without looking at the activity series, complete the following reactions, assuming aqueous media. If no reaction occurs, so indicate with NR.

a. $Li + ZnCl_2 \longrightarrow$ _____

b. $Sn + NaF \longrightarrow$ _____

c. $Ag + ZnI_2 \longrightarrow$ _____

d. $Mn + HCl \longrightarrow$ _____

e. $Cu + HI \longrightarrow$ _____

f. $Zn + H_2O(l) \longrightarrow$ _____

g. $K + AgCN \longrightarrow$ _____

h. $Pb + HBr \longrightarrow$ _____

i. $Fe + BaCl_2 \longrightarrow$ _____

j. $Sr + NiBr_2 \longrightarrow$ _____

8.10 You can use the activity series to determine the results of these reactions, again assuming aqueous solution.

a. $Ba + SrCl_2 \longrightarrow$ _____

b. $Cr + MgBr_2 \longrightarrow$ _____

c. $Ag + Au(CN)_3 \longrightarrow$ _____

d. $Pb + SnCl_2 \longrightarrow$ _____

OXIDIZING ACIDS

Since Cu is below H_2 in the activity series, we know that

$$Cu(s) + HCl(aq) \longrightarrow NR$$

This equation implies that Cu does not react with acids. However, such is not always the case, as can be seen from the reaction

$$3\ Cu(s) + 8\ HNO_3(aq) \longrightarrow 3\ Cu(NO_3)_2(aq) + 2\ NO(g) + 4\ H_2O$$

At first glance, it might appear that this reaction is a violation of the activity series. But it should be noted that Cu is *not* reducing the H^+ ion. Rather, the NO_3^- ion is oxidizing the Cu to Cu^{2+} and, correspondingly, N^{5+} is being reduced. An acid such as HNO_3 is sometimes referred to as an *oxidizing acid*. Reactions with oxidizing acids are not considered in the activity series. The activity series includes only elemental metals and H_2.

The most commonly discussed oxidizing acids are HNO_3 and hot, concentrated H_2SO_4. When Cu reacts with H_2SO_4(*hot, conc.*), the reaction is quite similar to the preceding one with HNO_3:

$$Cu(s) + 2\ H_2SO_4(h, c) \longrightarrow CuSO_4(aq) + SO_2(g) + 2\ H_2O$$

These two reactions, while appearing to be complex, are actually quite simple, and it is a relatively straightforward process to determine their products.

In the first reaction, HNO_3 oxidizes Cu^0 to Cu^{2+}. Remember that whenever an oxidation occurs, there must also be a reduction. In this case, the N is reduced from the $5+$ to the $2+$ state:

$$\overset{5+}{NO_3^-} \longrightarrow \overset{2+}{NO}$$

The same argument applies to the second reaction. Sulfuric acid oxidizes Cu^0 to Cu^{2+}, and the sulfur in the SO_4^{2-} must be reduced. Here we see

$$\overset{6+}{SO_4^{2-}} \longrightarrow \overset{4+}{SO_2}$$

In order to maintain electrical neutrality in the solution, the Cu^{2+} ion must, of course, have an anion to neutralize it, and the anion of the acid serves that purpose, giving, as a first step,

$$Cu + HNO_3 \longrightarrow Cu(NO_3)_2 + NO$$

and
$$Cu + H_2SO_4(h, c) \longrightarrow CuSO_4 + SO_2$$

Next we recognize that the H^+ ion cannot be reduced by Cu (activity series), so it combines with O^{2-} to form H_2O and we obtain

$$3\,Cu + 8\,HNO_3 \longrightarrow 3\,Cu(NO_3)_2 + 2\,NO + 4\,H_2O$$

and
$$Cu + 2\,H_2SO_4(h, c) \longrightarrow CuSO_4 + SO_2 + 2\,H_2O$$

Let's now reconsider the Cu and HCl system,

$$Cu + HCl \longrightarrow NR$$

We know that copper is below hydrogen in the activity series and therefore cannot reduce the H^+ ion, but inasmuch as Cu reacted with the NO_3^- and the SO_4^{2-} anions to form Cu^{2+}, we might ask why Cu doesn't react with the Cl^- ion to do the same thing. Recall our earlier discussion concerning oxidizing and reducing agents. For Cu to be oxidized, there must be an oxidizing agent. But the Cl^- ion cannot be an oxidizing agent. In order to do so it must itself be reduced. However, Cl^- is already in its lowest possible oxidation state and cannot be reduced further. Consequently, there is no way that copper can react with HCl.

_____ **EXERCISES** _____

8.11 Let's see if you've got it.

a. SO_4^{2-} and NO_3^- ions can act as oxidizing agents because _____.

b. Cl^-, Br^-, and I^- cannot act as oxidizing agents because _____.

8.12   Now give the results you would expect for these reactions.

a. $Ag + HNO_3(aq) \longrightarrow$ _____

b. $Cr + H_2SO_4(aq) \longrightarrow$ _____

c. $Hg + H_2SO_4(h, c) \longrightarrow$ _____

d. $Hg + H_2SO_4(aq) \longrightarrow$ _____

To complete our analysis of the reactions of oxidizing acids, consider the reaction of

$$Zn(s) + HNO_3(aq) \longrightarrow$$

Zinc is above H_2 in the activity series, so we would expect

$$Zn(s) + HNO_3(aq) \longrightarrow Zn(NO_3)_2(aq) + H_2(g)$$

and, to a limited extent, this reaction can occur. But remember, the NO_3^- ion is a very strong oxidizing agent, much stronger, in fact, than the H^+ ion. So other products might be formed. For instance, one might expect the same type of products as in the Cu-HNO_3 reaction. However, a glance at the activity series shows that Zn is a much better reducing agent than Cu. Consequently, Zn should be able to reduce nitrogen to even lower states than can Cu. Since, in the reaction between Cu and HNO_3, we saw that the nitrogen was reduced to $\overset{2+}{N}O$, it should not be surprising that Zn can reduce nitrogen to $\overset{1+}{N_2}O$, $\overset{0}{N_2}$, or even $\overset{3-}{N}H_3$. In fact, depending on the conditions, Zn can reduce the nitrogen all the way to the $3-$ state, giving

$$4\ Zn(s) + 9\ HNO_3(aq) \longrightarrow 4\ Zn(NO_3)_2(aq) + NH_3(aq) + 3\ H_2O$$

It is asking too much to expect you to predict the correct reduction products in each case. These often depend on such conditions as temperature, concentration, and pressure. For instance, in the previous reaction between Cu and HNO_3, the reduction product can, in fact, be NO_2 rather than NO. And, with the proper conditions, the above reaction between Zn and HNO_3 can give N_2 as the reduction product. Nevertheless, a reasonable guess can usually be made. If the products of a reaction are stated, you should be able to understand the reasons why they occur. Poor reducing agents such as the jewelry metals will tend to reduce the oxidizing agent to its next lower oxidation state and, sometimes, to the next state below that. With the stronger reducing agents, the lowest possible states might be expected. To further emphasize this point, let's look at these reactions:

$$\overset{0}{4 \text{ Mn}}(s) + 9 \overset{5+}{\text{HNO}_3}(aq) \longrightarrow 4 \overset{2+}{\text{Mn}}(\text{NO}_3)_2(aq) + \overset{3-}{\text{NH}_3}(aq) + 3 \text{ H}_2\text{O}$$

$$\overset{0}{\text{Hg}}(s) + 4 \overset{5+}{\text{HNO}_3}(aq) \longrightarrow \overset{2+}{\text{Hg}}(\text{NO}_3)_2(aq) + 2 \overset{4+}{\text{NO}_2}(g) + 2 \text{ H}_2\text{O}$$

From the activity series, we can see that Mn is a quite good reducing agent. In fact, it is above Zn in the series. Therefore, we would expect Mn to reduce $\text{NO}_3{}^-$ all the way to NH_3, the lowest possible oxidation state for nitrogen. By contrast, Hg is a poor reducing agent, and we would expect it to reduce nitrogen only to its next lower state, NO_2, in this case.

───────── **EXERCISE** ─────────

8.13 It's time for a challenge. Complete the following reactions:

a. $\text{Al} + \text{HNO}_3(aq) \longrightarrow$ _____

b. $\text{Sn} + \text{HNO}_3(aq) \longrightarrow$ _____

c. $\text{Sn} + \text{H}_2\text{SO}_4(h, c) \longrightarrow$ _____

d. $\text{Hg} + \text{H}_2\text{SO}_4(h, c) \longrightarrow$ _____

e. $\text{K} + \text{H}_2\text{SO}_4(h, c) \longrightarrow$ _____

f. $\text{Rb} + \text{H}_2\text{SO}_4(aq) \longrightarrow$ _____

════════════════════════

Before going on, a brief statement regarding terminology with respect to redox agents is in order. Consider the now familiar reaction between Cu and HNO_3. It is, in fact, the nitrogen in the $\text{NO}_3{}^-$ ion that is reduced in this reaction. Consequently, one would logically conclude that the $\overset{5+}{\text{N}}$ is the oxidizing agent, and this conclusion is certainly not incorrect. However, in practice, HNO_3 is much more commonly referred to as the oxidizing agent. This is because it is the chemical, nitric acid, that is actually added to the solution to serve this purpose. It is also true that the

NO_3^- ion can be considered to be the oxidizing agent. The point is that all three of these representations are correct, and you should be aware of all of them.

Oxidizing Anions.   There are a large number of anions that can act as oxidizing agents. The reduction products of these anions, however, can differ, depending on whether the solution is acidic or basic. We will consider only the acidic solutions here. Thus, these anions essentially amount to more examples of oxidizing acids.

We have already seen the reduction products of the NO_3^- ion. The same oxidation-reduction principles will now be extended to a number of other anions. In order to predict the results of a redox reaction with one of these ions, it is necessary to know the reduction products that might be formed. The following chart lists some of the likely products for a given oxidizing agent:

Oxidizing agent	Likely reduction products
$\overset{5+}{N}O_3^-$	$\overset{4+}{N}O_2 \quad \overset{2+}{N}O \quad \overset{1+}{N}_2O \quad \overset{0}{N}_2 \quad \overset{3-}{N}H_3$
$\overset{6+}{S}O_4^{2-}$	$\overset{4+}{S}O_2 \quad \overset{0}{S}_8 \quad \overset{2-}{H}_2S$
$\overset{7+}{C}lO_4^-$	$\overset{0}{C}l_2 \quad Cl^-$
$\overset{7+}{M}nO_4^-$	Mn^{2+}
$\overset{6+}{C}r_2O_7^{2-}, \overset{6+}{C}rO_4^{2-}$	Cr^{3+}

To illustrate the use of the chart, let's look at a specific reaction (unbalanced):

$$KMnO_4 + HCl \longrightarrow KCl + MnCl_2 + Cl_2 + H_2O$$

In analyzing this reaction, note that we have a strong oxidizing agent in the MnO_4^- ion. The K^+ ion and the H^+ ion cannot be oxidized because they are already in their highest oxidation states. The Cl^- ion, however, can be oxidized, and it is. Looking at the above chart, we see that the MnO_4^- ion would be expected to be reduced to Mn^{2+}. The oxidation product for the Cl^- ion is Cl_2, and this is the only logical product. Except for the $7+$ and the $5+$ states, the higher oxidation states of Cl are quite unstable, and the ClO_3^- ($5+$) and the ClO_4^- ($7+$) ions are excellent oxidizing agents themselves. Consequently, we would not expect the Cl^- ion to be oxidized to any higher state than observed in Cl_2. Finally, the H^+ ion combines with oxygen (O^{2-}) to form water, and additional Cl^- ions pair with the K^+ and the Mn^{2+} ions to maintain electrical neutrality.

EXERCISE

8.14   Try these. A number of different answers may be reasonable, so be sure yours is one of them.

a. $K_2Cr_2O_7(aq) + HCl(aq) \longrightarrow$ _____

b. $CuS(s) + HNO_3(aq) \longrightarrow$ _____

c. $S_8(s) + H_2SO_4(h, c) \longrightarrow$ _____

d. $H_2S(g) + HNO_3(aq) \longrightarrow$ _____

e. $KMnO_4(aq) + KBr(aq) + H_2SO_4(aq) \longrightarrow$ _____

STANDARD ELECTRODE POTENTIALS

Up until this point, we have focused on free metals (zero oxidation state) as reducing agents. Remember, however, that many metals have multiple positive oxidation states. Thus, their ions with the lower positive oxidation states can also act as reducing agents. For example, both Sn^{2+} and Sn^{4+} can act as oxidizing agents, with the Sn^{2+} being reduced to the free metal and the Sn^{4+} being reduced to either Sn^{2+} or Sn^0. But of these two, only the Sn^{2+} can act as a reducing agent, and it, of course, will be oxidized to Sn^{4+}. As an illustration consider the equation (not balanced)

$$Sn^{2+} + HNO_3 \longrightarrow Sn^{4+} + NO_2 + H_2O$$

A simpler type of reaction that illustrates the same point is

$$2\ Fe^{3+} + Sn^{2+} \longrightarrow 2\ Fe^{2+} + Sn^{4+}$$

or, in a more complete form, we might have

$$\overset{3+}{2\ FeCl_3} + \overset{2+}{SnCl_2} \longrightarrow \overset{2+}{2\ FeCl_2} + \overset{4+}{SnCl_4}$$

The question facing the student here is whether the reaction that actually takes place is

$$2\ Fe^{3+} + Sn^{2+} \longrightarrow 2\ Fe^{2+} + Sn^{4+}$$

or the reverse,

$$2 \ Fe^{2+} + Sn^{4+} \longrightarrow 2 \ Fe^{3+} + Sn^{2+}$$

In order to answer this question, we can refer to a table of *Standard Electrode Potentials* (see Table 8.2). This table represents a major extension of the activity series and lists the reducing abilities of ions and nonmetals as well as the metals we saw in the activity series. An actual table of Standard Electrode Potentials also lists numerical values of the reducing abilities of each species, but we will not be concerned with those in this discussion. It should be further pointed out that Table 8.2 is an abbreviated version of the table of Standard Electrode Potentials, and that all reactions are in aqueous media.

This listing may be a little confusing when compared to the activity series. The activity series lists the free metals in the order of their reducing abilities. In the course of a reaction, a metal will be oxidized and another metal ion will be reduced back to the metal. In the table of Standard Electrode Potentials, both the metal itself and the oxidized form of the metal are listed, with the oxidized form listed first. Just note that the order of the elements is still the same as in the activity series. Lithium is still the best reducing agent. But we can now see that the F^- ion is the worst reducing agent, which means that fluorine (F_2) is the best oxidizing agent.

Now we can return to the equation in question,

$$2 \ Fe^{3+} + Sn^{2+} \longrightarrow 2 \ Fe^{2+} + Sn^{4+}$$

From the list of Standard Electrode Potentials, we see that the $Sn^{4+} \rightarrow Sn^{2+}$ pair (these are referred to as a *couple*) is above the $Fe^{3+} \rightarrow Fe^{2+}$ couple, which means that Sn^{2+} is a better reducing agent than is Fe^{2+}. Observe that we are referring here to the ion with the lower oxidation state as the reducing agent. Since reducing agents are always oxidized, we can write

$$Sn^{2+} \longrightarrow Sn^{4+}$$

Further, reducing agents always reduce something when they react, and, in this case, the Sn^{2+} reduces the Fe^{3+}:

$$Fe^{3+} \longrightarrow Fe^{2+}$$

This finally gives us

$$2 \ Fe^{3+} + Sn^{2+} \longrightarrow 2 \ Fe^{2+} + Sn^{4+}$$

as the correct reaction.

Table 8.2   Standard Electrode Potentials

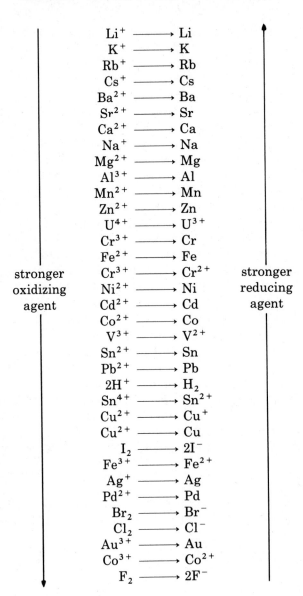

Now let's try a new one. Given the two couples

$$V^{3+} \longrightarrow V^{2+} \quad \text{and} \quad Cr^{3+} \longrightarrow Cr^{2+}$$

what reaction would you expect? The question, then, is whether

$$V^{3+} + Cr^{2+} \longrightarrow V^{2+} + Cr^{3+}$$

or
$$V^{2+} + Cr^{3+} \longrightarrow V^{3+} + Cr^{2+}$$

is more likely to occur. According to the listing of standard electrode potentials, the Cr couple is above the V couple. Consequently, Cr^{2+} is a better reducing agent than is V^{2+}. This means that Cr^{2+} can reduce V^{3+}, but V^{2+} cannot reduce Cr^{3+}, and the correct reaction therefore is

$$V^{3+} + Cr^{2+} \longrightarrow V^{2+} + Cr^{3+}$$

_____ **EXERCISES** _____

8.15 Here's your chance to show what you've learned. Given the redox couples, write down the reactions that do, in fact, occur.

a. $U^{4+} \longrightarrow U^{3+}$ and $Cr^{3+} \longrightarrow Cr^{2+}$ _____

b. $Cu^{2+} \longrightarrow Cu^+$ and $Fe^{3+} \longrightarrow Fe^{2+}$ _____

c. $Co^{3+} \longrightarrow Co^{2+}$ and $Sn^{4+} \longrightarrow Sn^{2+}$ _____

8.16 Try some nonmetals. The approach is the same as before.

a. $F_2 \longrightarrow 2\,F^-$ and $Cl_2 \longrightarrow 2\,Cl^-$ _____

b. $I_2 \longrightarrow 2\,I^-$ and $Br_2 \longrightarrow 2\,Br^-$ _____

c. $2\,H^+ \longrightarrow H_2$ and $Cl_2 \longrightarrow 2\,Cl^-$ _____

8.17 Now for an overall review. Give the products of the following reactions. If there is no reaction, so indicate with NR. Assume that all reactions are aqueous. Don't look at the activity series.

a. $Ba + ZnCl_2 \longrightarrow$ _____

b. $Pt + HCl \longrightarrow$ _____

c. $Ni + HBr \longrightarrow$ _____

d. $Ag + FeCl_3 \longrightarrow$ _____

e. $Na + H_2O(l) \longrightarrow$ _____

f. $Au + H_2O(g) \longrightarrow$ _____

You'll need the activity series for these.

g. $Na + CaCl_2 \longrightarrow$ _____

h. $Zn + MnBr_2 \longrightarrow$ _____

i. $Hg + AgCN \longrightarrow$ _____

j. $Mg + LiI \longrightarrow$ _____

Now try some oxidizing acids.

k. $Mg + HNO_3 \longrightarrow$ _____

l. $Pb + H_2SO_4(h, c) \longrightarrow$ _____

m. $Sn + HClO_4 \longrightarrow$ _____

n. $HNO_3 + H_2S \longrightarrow$ _____

o. $KMnO_4 + SnCl_2 + HCl \longrightarrow$ _____

Look at the table of Standard Electrode Potentials for these.

p. $Ag^+ + Fe^{2+} \longrightarrow$ _____

q. $Br_2 + Cl^- \longrightarrow$ _____

r. $Cr + H^+ \longrightarrow$ _____

s. $I^- + Cl_2 \longrightarrow$ _____

9

Some Important and Interesting Reactions

You have now worked your way through eight chapters of this book, and it is our hope that you have learned to predict the products of many types of chemical reactions. You now deserve a change of pace. In this, the final chapter, a number of specific and important reactions will be discussed. These will include reactions concerned with the production of industrial chemicals and metals, environmental problems such as acid rain, along with everyday reactions involving water softening and bread baking.

Chemistry textbooks have many chapters on descriptive chemistry, the chemistry of the elements. The study of this type of chemistry requires a large amount of reading and a familiarity with many reactions. In fact, learning descriptive chemistry can be rather trying. But learn it you should. Thus far, in your study of this book, you have been introduced to a variety of *type* reactions.

EXERCISE

9.1 See if you remember these. Tell the type of reaction in each case.

a. $2 \text{ Na}(s) + \text{Cl}_2(g) \longrightarrow 2 \text{ NaCl}(s)$ _____

b. $\text{Zn}(s) + 2 \text{ HCl}(aq) \longrightarrow \text{ZnCl}_2(aq) + \text{H}_2(g)$ _____

c. $\text{BaCl}_2(aq) + \text{Na}_2\text{SO}_4(aq) \longrightarrow \text{BaSO}_4(s) + 2 \text{ NaCl}(aq)$ _____

d. $\text{CaCO}_3(s) \xrightarrow{\Delta} \text{CaO}(s) + \text{CO}_2(g)$ _____

e. $SO_2(g) + H_2O \longrightarrow H_2SO_3(aq)$ _____

f. $BaO(s) + H_2O \longrightarrow Ba(OH)_2(aq)$ _____

g. $HCl(aq) + NaOH(aq) \longrightarrow NaCl(aq) + H_2O$ _____

h. $HCl(g) + H_2O \longrightarrow H_3O^+(aq) + Cl^-(aq)$ _____

i. $H^+(aq) + NH_3(aq) \longrightarrow NH_4^+(aq)$ _____

j. $Ag^+(aq) + 2\ NH_3(aq) \longrightarrow [Ag(NH_3)_2]^+(aq)$ _____

As you study descriptive chemistry, look for these type reactions. You will find them recurring many times. As you will come to see, an awareness of the various type reactions will prove to be invaluable in your study of descriptive chemistry. And to help you accomplish this is, of course, one of the primary purposes of this book.

AN APOCRYPHAL STORY—ONE THAT COULD HAVE HAPPENED

But if it didn't happen exactly this way, you will still learn how to write the equations! The alchemists were obsessed with making gold, and after repeated failures, some of them got the idea that the secret lay in first dissolving it. Since no known solvents would dissolve gold, they began to produce new compounds in a search for a solvent for gold.

Sulfuric acid was discovered in the sixteenth century by heating *green vitriol*, a hydrated iron sulfate.

$$FeSO_4 \cdot H_2O(s) \xrightarrow{\Delta} H_2SO_4(l) + FeO(s)$$

But it would not dissolve gold. The sulfuric acid was then treated with sodium chloride to produce a new acid, hydrochloric acid,

$$2\ NaCl(s) + H_2SO_4(l) \xrightarrow{\Delta} Na_2SO_4(s) + 2\ HCl(g)$$

This reaction was first carried out by Johann Rudolph Glauber, a noted alchemist. In fact, to this day Na_2SO_4 is called *Glauber's salt* and it is now used in the production of brown paper and corrugated boxes. But neither the HCl nor the H_2SO_4 will dissolve gold.

Nitric acid (HNO_3) was produced by a similar process,

$$2\ NaNO_3(s) + H_2SO_4(l) \xrightarrow{\Delta} Na_2SO_4(s) + 2\ HNO_3(g)$$

Note that in the formation of both hydrochloric acid and nitric acid, the salt of the acid reacted with sulfuric acid to form the new acid. None of these acids dissolves gold. Nitric acid, however, being a particularly good oxidizing agent, dissolves all the familiar metals except gold and platinum.

_____ **EXERCISE** _____

9.2 Nitric acid will oxidize and, thereby, dissolve the following metals. Which ions would you expect to be formed?

a. Cu $\longrightarrow$ _____   **b.** Ag $\longrightarrow$ _____

c. Al $\longrightarrow$ _____   **d.** Zn $\longrightarrow$ _____

e. Why can't Cl^- be an oxidizing agent? _____

Recall that gold is at the bottom of the activity series which means that it has a tendency not to oxidize and, therefore, a tendency not to dissolve. Gold does dissolve, however, in a 3:1 volume mixture of the concentrated acids, HCl and HNO_3.

$$Au + 4\,H^+ + 4\,Cl^- + NO_3^- \longrightarrow [AuCl_4]^- + NO + 2\,H_2O$$

Gold is oxidized by the nitric acid to the 3+ oxidation state, but, because of the instability of this state, the Au^{3+} can only exist as a complex ion. The Cl^- ion from the HCl then acts as the complexing agent. Thus in order to dissolve gold, both an oxidizing agent and a complexing agent are needed. Imagine the excitement of the alchemists when a solvent was finally found. The name _aqua regia_ (royal water) seemed quite appropriate. They surely must have felt at that time that they were on the threshold of fulfilling their dreams. Such dreams, as we now know, were not to be realized.

_____ **EXERCISE** _____

9.3 When gold dissolves in aqua regia, the formula of the gold complex

ion is _____, and it is named _____.

SULFURIC ACID

As the need for sulfuric acid increased and the supply of green vitriol decreased, it became imperative that a new method of production be discovered, one that would meet the needs for sulfuric acid and, at the same time, not depend on the dwindling supply of green vitriol.

It was found that when sulfur is burned in a glass container in the presence of $NaNO_3$ and H_2O, sulfuric acid is produced. This method evolved into the *lead chamber process* which permitted the mass production of H_2SO_4. The chemistry of this process is very complicated and, even today, poorly understood. A stream of gas containing SO_2, O_2, NO, NO_2, and steam is fed into large lead-lined chambers, hence the name lead chamber process. The overall reaction is

$$2 H_2O + 2 SO_2(g) + O_2(g) \longrightarrow 2 H_2SO_4(l)$$

but what takes place in between is much more complicated. A simplified version is

$$2 NO(g) + O_2(g) \longrightarrow 2 NO_2(g)$$

$$SO_2(g) + NO_2(g) \longrightarrow SO_3(g) + NO(g)$$

$$SO_3(g) + H_2O \longrightarrow H_2SO_4(l)$$

Today the demand for sulfuric acid has become so large that it is now the world's number one industrial chemical. At the present time, most of the commercially prepared H_2SO_4 is made by the *contact method*. In this process, SO_2 reacts with O_2 in the presence of a V_2O_5 catalyst. Thus, we have simply

$$2 SO_2(g) + O_2(g) \xrightarrow{V_2O_5} 2 SO_3(g)$$

Based on what we have learned up to this point, it would seem that the SO_3 should be simply mixed with H_2O to give the H_2SO_4. This, however, is not the case, because SO_3 does not readily dissolve in H_2O. Consequently, the rate of production of sulfuric acid can be increased by dissolving the SO_3 in a concentrated solution of H_2SO_4, giving fuming (or pyro) sulfuric acid, $H_2S_2O_7$. This is then diluted with H_2O to give the desired concentration of H_2SO_4:

$$SO_3(g) + H_2SO_4(l) \longrightarrow H_2S_2O_7(l)$$

and $\qquad\qquad H_2S_2O_7(l) + H_2O \longrightarrow 2 H_2SO_4(l)$

NITRIC ACID

Today nitric acid is produced by a method that uses only ammonia and air as the raw materials. This conversion was first made possible on a commercial level with the development of the *Haber process* shortly before World War I,[1] a process that permitted the commercial preparation of NH_3 from its elements. In the Haber process we have

$$N_2(g) + 3\ H_2(g) \longrightarrow 2\ NH_3(g)$$

The preparation of HNO_3 is then accomplished with the *Ostwald process*, which involves the following three steps:

$$4\ NH_3(g) + 5\ O_2(g) \longrightarrow 4\ NO(g) + 6\ H_2O(g)$$

The gases are cooled and more air is added, resulting in the reaction

$$2\ NO(g) + O_2(g) \longrightarrow 2\ NO_2(g)$$

The NO_2 then reacts with H_2O to form the HNO_3:

$$3\ NO_2(g) + H_2O \longrightarrow 2\ HNO_3(aq) + NO(g)$$

Finally, the NO is recycled and oxidized to form NO_2, and the process is repeated.

It is important to note that, although NO_2 reacts with H_2O to form HNO_3, it is not the acid anhydride of HNO_3. Remember that an acid anhydride reacts with H_2O to give the acid and nothing more. Further, there should be no changes in oxidation states.

––––––––––– **EXERCISE** –––––––––––

9.4 Try this for a review. Give the products of the indicated reactions and the oxidation states of the central atoms.

a. $SO_2(g) + H_2O \longrightarrow$ _____H_2SO_3_____

S in SO_2 _4+_ ; S in product _4+_

b. $CO_2(g) + H_2O \longrightarrow$ _____

C in CO_2 _____ ; C in product _____

––––––––––––––––––––––––––––––––––––

[1] It is quite difficult to get nitrogen and hydrogen to react, as very special conditions of temperature and pressure are needed. See any general chemistry text for details.

c. $SO_3(g) + H_2O \longrightarrow$ _____

S in SO_3 _____ ; S in product _____

d. $P_4O_6(s) + H_2O \longrightarrow$ _____

P in P_4O_6 _____ ; P in product _____

e. $P_4O_{10}(s) + H_2O \longrightarrow$ _____

P in P_4O_{10} _____ ; P in product _____

It might be interesting to spend some time here analyzing the reaction

$$3\ NO_2(g) + H_2O \longrightarrow 2\ HNO_3(aq) + NO(g)$$

to see if there is some systematic way we could have attempted the prediction of these products. Again, we must realize that a prediction does not always give the correct answer. Reasonable guesses, however, are better than wild guesses!

First note that $4+$ is not a common oxidation state for nitrogen. In fact, except for the oxide, NO_2, you have not seen nitrogen with a $4+$ oxidation state. Thus, if a reaction does occur, it must be a redox reaction in which the nitrogen is oxidized, reduced, or both. If we first imagine the N to be oxidized, it can only go to the $5+$ oxidation state, the highest possible oxidation state for nitrogen. But if nitrogen is oxidized, something must be reduced. The possible choices are

N: $3+, 2+, 1+, 0, 3-$
H: $0, 1-$
O: no choices; it is already in its lowest oxidation state.

The possible reactions then become

$$NO_2 + H_2O \longrightarrow HNO_3 + \overset{3+}{HNO_2}$$

$$NO_2 + H_2O \longrightarrow HNO_3 + \overset{2+}{NO}$$

$$NO_2 + H_2O \longrightarrow HNO_3 + \overset{1+}{N_2O}$$

$$NO_2 + H_2O \longrightarrow HNO_3 + \overset{0}{N_2}$$

$$NO_2 + H_2O \longrightarrow HNO_3 + \overset{3-}{NH_3}$$

$$NO_2 + H_2O \longrightarrow HNO_3 + H_2$$

The last one of these should readily be discarded. The reduction of the H^+ ion implies that NO_2 is a reasonably good reducing agent. However, in NO_2, N is in its next to highest oxidation state. It is much more reasonable to consider it to be an oxidizing agent. In addition, if the reaction did occur as written, the HNO_3 would oxidize the H_2, pushing the reaction back to the left. This leaves the first five reactions as possibilities, and the choice among these is not straightforward. In fact, all of these reactions are reasonable. About the only way we could come up with the correct result would be to use a table of Standard Electrode Potentials, one much more complete than the one given in Chapter 8. We might narrow the possibilities somewhat by knowing that disproportionation reactions (reactions in which the same element is both oxidized and reduced) do not usually show large differences in oxidation states. On this basis, we could reasonably guess that N_2 and NH_3 should be eliminated as possible products.

You might wonder why we don't get N_2O_3 rather than HNO_2 for the $3+$ oxidation state of nitrogen. The reason is simply that N_2O_3 would react with H_2O to form HNO_2; that is, N_2O_3 is the acid anhydride of HNO_2.

We have considered the possibility that the nitrogen is oxidized. Now let's consider the possibilities if the nitrogen is reduced. This will require that something be oxidized, and our choices are

N: 5+

O: 0

H: no choices; it is already in its highest oxidation state.

The use of the $5+$ state for N offers nothing new. We have already considered all of these possibilities in the previous reactions. However, the oxidation of O^{2-} to O_2 is new. For instance,

$$NO_2(g) + H_2O \longrightarrow HNO_2(l) + O_2(g)$$

The problem with this reaction is that O_2 is a very good oxidizing agent, and the reverse reaction would be the more likely. Putting it another way, one might conclude that O^{2-} is not a sufficiently strong reducing agent to reduce NO_2. But again, to be confident of the results, we would have to use a table of Standard Electrode Potentials.

In conclusion we can only say that the first five of these reactions are all reasonable, and, in fact, any one of them might occur with the proper conditions of temperature and pressure.

---------   **EXERCISE**   ---------

9.5   Now let's see if you can give the correct products for the following reactants.

a. $NaCl(s) + H_2SO_4(l) \xrightarrow{\Delta}$ _____

b. $Ag(s) + HNO_3(aq) \longrightarrow$ _____

c. $Au(s) + H^+(aq) + Cl^-(aq) + NO_3^-(aq) \longrightarrow$ _____

d. $NO_2(g) + SO_2(g) \longrightarrow$ _____

e. $SO_3(g) + H_2O \longrightarrow$ _____

f. $NH_3(g) + O_2(g) \xrightarrow[500°-1000°]{Pt-Rh}$ _____

g. $NO(g) + O_2(g) \longrightarrow$ _____

h. $NO_2(g) + H_2O \longrightarrow$ _____

METALLURGY

Although metallurgy involves virtually all aspects of the study of metals, it is commonly identified with the obtaining of metals from their ores. We will focus on only a few of the interesting and important separations.

It is no accident that the first metals that seem to have been used by the ancients were gold and copper. Both of these are near the bottom of the activity series. Because of this, they are among the very few metals that occur in nature as the pure elements. The first working of metals was recorded in the Mediterranean nations of Egypt and Mesopotamia. In this region, objects of both copper and gold have been dated to as early as 3000–4000 B.C. They were highly prized because they could be worked and hammered into various forms.

Copper.   One of the truly great steps in the advancement of civilization came with the recovery of copper from its ores. The supply of free copper was quite limited. In fact, at that time the amount was probably so small that it could only be used for ornamentation. We can reasonably assume that certain kinds of rocks accidentally came in contact with fire. These rocks, known today as copper ores, yielded copper when heated. We can imagine the excitement when it was realized that by simply heating these rocks the copper supply could be increased many fold.

Ores are minerals from which metals may be profitably extracted. It is likely that the first copper ore to be discovered was malachite [$CuCO_3 \cdot Cu(OH)_2$]. This ore will yield copper in a wood fire where the charcoal plays an important role. The reaction is

$$\text{malachite} + C(s) \xrightarrow{\Delta} Cu(s) + CO_2(g) + H_2O$$

This reaction may seem rather confusing at first, but consider the products you would expect for

$$CuCO_3(s) \xrightarrow{\Delta}$$
$$Cu(OH)_2(s) \xrightarrow{\Delta}$$

In both cases CuO would be expected along with CO_2 from the decomposition of $CuCO_3$, and H_2O from the decomposition of $Cu(OH)_2$. Since $CuCO_3$ and $Cu(OH)_2$ can be thought of as the ingredients, in some form, of malachite, we might now imagine the reaction to be

$$2\ CuO(s) + C(s) \xrightarrow{\Delta} 2\ Cu(s) + CO_2(g)$$

Charcoal (C) is a very good reducing agent at high temperatures and is used in many metallurgical processes.

Another common copper ore is chalcocite (Cu_2S). Although the actual process is quite involved, the reduction of the ore can be expressed with two equations

$$2\ Cu_2S(s) + 3\ O_2(g) \xrightarrow{\Delta} 2\ Cu_2O(s) + 2\ SO_2(g)$$
$$2\ Cu_2O(s + Cu_2S(s) \xrightarrow{\Delta} 6\ Cu(s) + SO_2(g)$$

Zinc, Lead and Tin. The sulfides are common ores of zinc and lead, sphalerite (ZnS) and galena (PbS). The common ore of tin is cassiterite (SnO_2). The first step in the recovery of the metal from the sulfide ores is *roasting* in air, giving

$$2\ ZnS(s) + 3\ O_2(g) \xrightarrow{\Delta} 2\ ZnO(s) + 2\ SO_2(g)$$

and

$$2\ PbS(s) + 3\ O_2(g) \xrightarrow{\Delta} 2\ PbO(s) + 2\ SO_2(g)$$

The next step is common for all three metals:

$$ZnO(s) + C(s) \xrightarrow{\Delta} Zn(l) + CO(g)$$

$$PbO(s) + C(s) \xrightarrow{\Delta} Pb(l) + CO(g)$$

$$SnO_2(s) + 2\ C(s) \xrightarrow{\Delta} Sn(l) + 2\ CO(g)$$

Note first that the roasting of sulfide ores normally gives the metal oxide. Then note the use of charcoal as a reducing agent. It should also be noted that CO can act as a reducing agent. Thus we can have

$$PbO(s) + CO(g) \xrightarrow{\Delta} Pb(s) + CO_2(g)$$

All three of these metals have been extremely important in the development of civilization. It was the forming of an alloy of copper and tin that gave us the bronze age. Although it is not known how the discovery of bronze actually came about, it probably resulted from the accidental mixing of copper and tin ores on a wood fire.

_____ **EXERCISE** _____

9.6 Give the products for these.

a. $Cu_2S(s) + O_2(g) \xrightarrow{\Delta}$ _____

b. $NO_2(g) + H_2O \longrightarrow$ _____

c. $SnO_2(s) + C(s) \xrightarrow{\Delta}$ _____

d. $NaNO_3(s) + H_2SO_4(l) \xrightarrow{\Delta}$ _____

e. $CaCO_3(s) \xrightarrow{\Delta}$ _____

f. $PbS(s) + O_2(g) \xrightarrow{\Delta}$ _____

g. $Cu_2S(s) + Cu_2O(s) \xrightarrow{\Delta}$ _____

h. $ZnO(s) + C(s) \xrightarrow{\Delta}$ _____

i. $SO_2(g) + H_2O \longrightarrow$ _____

j. $PbO(s) + CO(g) \xrightarrow{\Delta}$ _____

Silver and Gold. Much of the silver used in this country is obtained electrolytically as a by-product of the refining of copper. And we are all familiar with the use of panning and sluicing for gold as a result of our exposure to the movies and stories of the gold-rush days. These stories imply that gold tends to be found in the pure state, and it does. But when working with low grade ores, the convenience and the romance end, and more sophisticated techniques are required. As it turns out, the same approach is used for both gold and silver to recover the elements from their low-grade ores.

Remember that aqua regia dissolves gold by forming an $AuCl_4^-$ complex ion. An analogous approach is used to obtain both silver and gold from their low-grade ores. But here the CN^- complex ion is used:

$$4 \text{ Ag}(s) + 8 \text{ CN}^-(aq) + O_2(g) + 2 \text{ H}_2O \longrightarrow$$
$$4 \text{ Ag(CN)}_2^-(aq) + 4 \text{ OH}^-(aq)$$

and

$$4 \text{ Au}(s) + 8 \text{ CN}^-(aq) + O_2(g) + 2 \text{ H}_2O \longrightarrow$$
$$4 \text{ Au(CN)}_2^-(aq) + 4 \text{ OH}^-(aq)$$

Silver frequently occurs as the sulfide (Ag_2S) and the chloride (AgCl). These compounds will also react to form the cyanide complex ion. The final step in the recovery of the metal is reduction of the complex ion with Zn,

$$2 \text{ Ag(CN)}_2^-(aq) + \text{Zn}(s) \longrightarrow 2 \text{ Ag}(s) + \text{Zn(CN)}_4^{2-}(aq)$$
$$2 \text{ Au(CN)}_2^-(aq) + \text{Zn}(s) \longrightarrow 2 \text{ Au}(s) + \text{Zn(CN)}_4^{2-}(aq)$$

From a chemical standpoint, the whole procedure is quite simple. The ore is crushed and an aqueous solution of NaCN is added. The cyanide complex of the gold or silver, or both, is formed. These complexes are water soluble and can be separated from the rock. Zinc is then added to the final solution, and the Ag^+ and Au^+ ions are reduced.

EXERCISES

9.7   In the reaction between $Ag(CN)_2^-$ and Zn,

a. Zn is the _____ agent.

b. _____ gives up electrons.

c. _____ is oxidized.

9.8   Name the following ions.

a. $Ag(CN)_2^-$ _____.

b. $AuCl_4^-$ _____.

c. $Zn(CN)_4^{2-}$ _____.

9.9   Now give these a try. Write the products.

a. $Cu_2S(s) + O_2(g) \xrightarrow{\Delta}$ _____

b. $Zn(s) + Ag(CN)_2^-(aq) \longrightarrow$ _____

c. $NaI(s) + H_2SO_4(l) \xrightarrow{\Delta}$ _____

d. $Au(s) + HNO_3(aq) + HCl(aq) \longrightarrow$ _____

e. $Cu(s) + HNO_3(aq) \longrightarrow$ _____

_____

Iron.   There is considerable evidence that iron was well known in the Middle East as early as 2500 B.C. However, it was not generally used for tools and weapons until much later. The principal source of native iron was probably meteorites, and these would not offer a sufficient amount for uses beyond ornamentation. But the reduction of iron from its ores was certainly accomplished by 1500 B.C., most likely by the Hittites or the Assyrians. Both of these peoples were well known for their work with iron, and with this work they helped lead to another major advance in civilization—the *iron age.*

Iron is obtained chiefly from the ore hematite (Fe_2O_3). This ore cannot be reduced merely with a wood fire as was the case with the copper ore malechite. Wood simply does not provide sufficient heat. However, using charcoal and bellows, one can cause the following reactions to occur:

$$2 \text{ C (Charcoal)} + O_2(g) \xrightarrow{\Delta} 2 \text{ CO}(g)$$

$$3 \text{ Fe}_2O_3(s) + CO(g) \xrightarrow{\Delta} 2 \text{ Fe}_3O_4(s) + CO_2(g)$$

$$\text{Fe}_3O_4(s) + CO(g) \xrightarrow{\Delta} 3 \text{ FeO}(s) + CO_2(g)$$

$$\text{FeO}(s) + CO(g) \xrightarrow{\Delta} \text{Fe}(s) + CO_2(g)$$

Note that the iron is reduced in steps by the carbon monoxide.

—————— **EXERCISE** ——————

9.10 What is the oxidation state of Fe in

a. Fe_2O_3 _____   **b.** Fe_3O_4 _____

c. FeO _____

——————————————————

If the last three of the above equations are added up, the net result is the equation

$$Fe_2O_3(s) + 3\ CO(g) \xrightarrow{\ \Delta\ } 2\ Fe(s) + 3\ CO_2(g)$$

In the present-day production of iron, limestone ($CaCO_3$) is also added to the system to separate the ever-present impurities from the iron. These are most commonly silica (SiO_2) and alumina (Al_2O_3). The limestone is decomposed by the heat to form lime (CaO):

$$CaCO_3(s) \xrightarrow{\ \Delta\ } CaO(s) + CO_2(g)$$

The lime then reacts with SiO_2 and Al_2O_3 to form slag, which is less dense than the molten iron and can easily be removed:

$$CaO(s) + SiO_2(s) \xrightarrow{\ \Delta\ } CaSiO_3(s)$$

and $\qquad CaO(s) + Al_2O_3(s) \xrightarrow{\ \Delta\ } Ca(AlO_2)_2(s)$

It is both interesting and helpful to analyze these two reactions. To begin with, they are acid-base reactions. You saw a similar type of reaction in Chapter 7, namely

$$Na_2O(s) + SO_3(g) \longrightarrow Na_2SO_4(s)$$

Na_2O is a basic anhydride and SO_3 is an acid anhydride. Likewise, CaO is a basic anhydride and, in this case, both SiO_2 and Al_2O_3 act as acid anhydrides.

—————— **EXERCISE** ——————

9.11 Although SiO_2 does not actually react with H_2O, assume that it does, and give the *meta-* forms of the expected acids for these two reactions.

a. $SiO_2(s) + H_2O \longrightarrow$ _____

b. $Al_2O_3(s) + H_2O \longrightarrow$ _____

_____

If we look at the same type of reaction with CaO, we obtain

$$CaO(s) + H_2O \longrightarrow Ca(OH)_2(s)$$

Now we can consider the previous reactions in terms of simple protonic acid–hydroxyl base type reactions:

$$Ca(OH)_2(s) + H_2SiO_3(aq) \longrightarrow CaSiO_3(s) + H_2O$$

and

$$Ca(OH)_2(s) + 2\ HAlO_2(aq) \longrightarrow Ca(AlO_2)_2(s) + 2\ H_2O$$

As was pointed out in Chapter 7, when an acidic oxide and a basic oxide react, the salt is the same as the one formed by the reaction of the corresponding protonic acid and hydroxyl base but, of course, no H_2O is formed.

To complete our discussion of iron, it should be mentioned that the triumph of the iron age over the bronze age resulted from the fact that iron tools are harder and hold their edges better. That is, in a battle, an iron sword is superior to a bronze sword. But in actuality, iron is not such a hard metal. It is steel that is hard, and it was steel that gave birth to the iron age. Steel is an alloy of iron and a small amount of carbon. It is reasonable to assume that, at least in the beginning, the formation of steel was accidental, resulting from the contamination of the iron by the charcoal used in its production.

_____ **EXERCISE** _____

9.12 Here are a few more to keep you from forgetting what you have learned. Give the products for the following reactions.

a. $Fe_2O_3(s) + CO(g) \xrightarrow{\Delta}$ _____

b. $CaO(s) + SiO_2(s) \xrightarrow{\Delta}$ _____

c. $NaBr(s) + H_2SO_4(l) \xrightarrow{\Delta}$ _____

d. $Zn(s) + HNO_3(aq) \longrightarrow$ _____

e. $Ag(CN)_2{}^-(aq) + Zn(s) \longrightarrow$ _____

f. $FeO(s) + CO(g) \xrightarrow{\Delta}$ _____

g. $N_2O_3(g) + H_2O \longrightarrow$ _____

h. $NO_2(g) + H_2O \longrightarrow$ _____

i. $H_2(g) + N_2(g) \xrightarrow{\Delta}$ _____

j. $SnO_2(s) + C(s) \xrightarrow{\Delta}$ _____

CALCIUM AND SODIUM CARBONATES

Calcium carbonate ($CaCO_3$) is one of the earth's most abundant minerals. It is found in nature as the major constituent of limestone, marble, chalk, pearls, and seashells. We have already seen that it decomposes on heating as follows:

$$CaCO_3(s) \xrightarrow{\Delta} CaO(s) + CO_2(g)$$

CaO is known as quicklime and has many industrial uses. We have seen its use in the metallurgy of iron. It is also used in the making of glass. Portland cement contains roughly 60% CaO, and mortar for brick laying contains additional quicklime. In fact, a good brick mortar can be made simply from quicklime, sand, and water. The addition of H_2O to the quicklime results in *slaked lime* which is hydrated $Ca(OH)_2$. The slaked lime then gradually reacts with CO_2 from the atmosphere to form $CaCO_3$, thereby bonding the sand (SiO_2) in the mortar.

_____ **EXERCISES** _____

9.13 Complete the following reactions which are important in the making and setting of mortar.

a. $CaCO_3(s) \xrightarrow{\Delta}$ _____

b. $CaO(s) + H_2O \longrightarrow$ _____

c. $Ca(OH)_2(s) + CO_2(g) \longrightarrow$ _____

9.14 CaO is the _____ _____ of $Ca(OH)_2$, and CO_2 is the acid anhydride of _____ .

Sodium carbonate (Na_2CO_3), which is commonly called soda ash, and sodium hydrogen carbonate ($NaHCO_3$), the familiar baking soda, are both produced by the *Solvay process*. This process begins with the decomposition of limestone:

$$CaCO_3(s) \xrightarrow{\Delta} CaO(s) + CO_2(g)$$

The CO_2, along with NH_3, is bubbled into a concentrated solution of NaCl. The CO_2 reacts with the H_2O to form carbonic acid,

$$CO_2(g) + H_2O \longrightarrow H_2CO_3(aq)$$

and the H_2CO_3 reacts with NH_3, giving

$$H_2CO_3(aq) + NH_3(g) \longrightarrow NH_4{}^+(aq) + HCO_3{}^-(aq)$$

_____ **EXERCISE** _____

9.15 In the preceding reaction H_2CO_3 is a protonic _____ and NH_3 is a protonic _____ .

Sodium hydrogen carbonate ($NaHCO_3$) is less soluble than NaCl. Consequently, as the concentration of the $HCO_3{}^-$ ion increases, it combines with the Na^+ ion in the salt solution and selectively precipitates as $NaHCO_3$. That is,

$$NH_4{}^+(aq) + Cl^-(aq) + Na^+(aq) + HCO_3{}^-(aq) \longrightarrow$$
$$NaHCO_3(s) + NH_4{}^+(aq) + Cl^-(aq)$$

The $NaHCO_3$ can then be converted to Na_2CO_3 by heating:

$$2\,NaHCO_3(s) \xrightarrow{\Delta} Na_2CO_3(s) + H_2O(g) + CO_2(g)$$

Interlude. Let's return to our discussion about reasonable guesses when determining the products of a reaction. When compounds containing the carbonate ion ($CO_3{}^{2-}$) are decomposed, CO_2 will be a product, and compounds containing the hydrogen carbonate ion ($HCO_3{}^-$) will decompose to give both CO_2 and H_2O. Furthermore, compounds containing C and H or C, H, and O will yield CO_2 and H_2O when they react with O_2.

EXERCISE

9.16 Make reasonable guesses of the products of the following reactions.

a. $CaCO_3(s) \xrightarrow{\Delta}$ _____ + _____

b. $H_2O_2(aq) \longrightarrow O_2(g) +$ _____

c. $H_2CO_3(aq) \rightleftharpoons$ _____ + _____

d. $NaHCO_3(s) \xrightarrow{\Delta}$ _____ + _____ + _____

e. $C_2H_6(g) + O_2(g) \xrightarrow{\Delta}$ _____ + _____

f. $C_2H_5OH(l) + O_2(g) \xrightarrow{\Delta}$ _____ + _____

Let's look again at the Solvay process. The remaining solution containing the NH_4Cl is now treated with the CaO left over from the decomposition of the $CaCO_3$. The reactions that occur are

$$CaO(s) + H_2O \longrightarrow Ca(OH)_2(aq)$$

and

$$Ca(OH)_2(s) + 2\ NH_4Cl(aq) \longrightarrow CaCl_2(aq) + 2\ H_2O + 2\ NH_3(g)$$

The regenerated ammonia is then used over again and again in the process to produce more Na_2CO_3.

Extremely large amounts of Na_2CO_3 are produced each year. It is used in the manufacture of glass, paper, soap, and detergents, and as an alkali in other chemical industries. Additionally, it can be used as a water softening agent.

Sodium hydrogen carbonate ($NaHCO_3$) is found in nearly every household under the name of baking soda. A familiar *kitchen chemistry* reaction results from the addition of vinegar (acetic acid, CH_3COOH, abbreviated as HOAc) to baking soda. The fizzing that occurs is due to the release of CO_2:

$$NaHCO_3(aq) + HOAc(aq) \longrightarrow NaOAc(aq) + H_2CO_3(aq)$$

EXERCISES

9.17 But we know that H_2CO_3 decomposes as follows:

$$H_2CO_3(aq) \longrightarrow \text{_____} + \text{_____}$$

9.18 Thus, we can say

$$NaHCO_3(aq) + HOAc(aq) \longrightarrow NaOAc(aq) + \underline{\hspace{2cm}} + \underline{\hspace{2cm}}$$

When baking soda is added to dough, it is necessary to add an acid. This usually takes the form of buttermilk, vinegar, or lemon juice. In ionic form, we can represent the reaction as

$$NaHCO_3(aq) + H^+(aq) \longrightarrow Na^+(aq) + CO_2(g) + H_2O$$

It is the bubbles of CO_2 that cause the dough to rise, thus making the baked product light.

The hydrogen carbonate ion is particularly interesting in that it can act as either a protonic acid or a protonic base, as can be seen from the reactions

$$HCO_3^-(aq) + H^+(aq) \longrightarrow H_2CO_3(aq)$$

and $$HCO_3^-(aq) + OH^-(aq) \longrightarrow H_2O + CO_3^{2-}(aq)$$

It is for this reason that baking soda is used to neutralize both acid and base spills in the laboratory.

HARD WATER

Hard water usually results from the presence of Ca^{2+}, Mg^{2+}, and Fe^{2+} ions. These ions react with the long-chain organic salts that characterize soaps to form precipitates that show up as the scum in pots and pans and the ring found in the bathtub. The general form of the reaction, in which M stands for a metal, is[2]

$$M^{2+}(aq) + 2\ Na^+(aq) + 2\ C_{17}H_{35}COO^-(aq) \longrightarrow$$
$$2\ Na^+(aq) + M^{2+}(C_{17}H_{35}COO^-)_2(s)$$

The choices one has are (a) live with the scum and use more soap, (b) use synthetic detergents, or (c) soften the water. With regard to the latter approach, there are two types of hard water. The first is associated with the presence of the hydrogen carbonates of these ions and is referred to as *temporary* or *carbonate* hardness. It can be softened merely by boiling.

[2] Note that $Na^+(aq) + C_{17}H_{35}COO^-(aq)$ is the dissolved soap.

Thus,

$$Ca(HCO_3)_2(aq) \xrightarrow{\Delta} CaCO_3(s) + H_2O + CO_2(g)$$

In contrast, if the anion is that of a soluble salt of the Ca^{2+}, Mg^{2+}, or Fe^{2+} ions, such as NO_3^- or Cl^-, the result will be *permanent* hard water. Permanent hard water can be softened in a number of ways, one being the addition of Na_2CO_3. This commonly takes the form of washing soda ($Na_2CO_3 \cdot 10 H_2O$), a product that can be purchased in the grocery store. The carbonate ion precipitates the offending ions, leaving the Na^+ ion as the cation in the solution. Using the Ca^{2+} ion for illustration, the reaction is

$$Ca^{2+}(aq) + Na_2CO_3(aq) \longrightarrow CaCO_3(s) + 2 Na^+(aq)$$

———————— **EXERCISE** ————————

9.19 Now for some more review.

a. $CaO(s) + H_2O \longrightarrow$ _____

b. $Ca(OH)_2(aq) + CO_2(g) \longrightarrow$ _____

c. $NaHCO_3(s) \xrightarrow{\Delta}$ _____

d. $HCO_3^-(aq) + H^+(aq) \longrightarrow$ _____

e. $Au(s) + HCl(aq) + HNO_3(aq) \longrightarrow$ _____

f. $H_2O + SO_3(g) \longrightarrow$ _____

g. $NO_2(g) + H_2O \longrightarrow$ _____

h. $SnO_2(s) + C(s) \xrightarrow{\Delta}$ _____

i. $Cu_2S(s) + O_2(g) \xrightarrow{\Delta}$ _____

j. $AgCl(s) + Cl^-(aq) \longrightarrow$ _____

k. $Fe_2O_3(s) + CO(g) \xrightarrow{\Delta}$ _____

l. $CaO(s) + SiO_2(s) \xrightarrow{\Delta}$ _____

m. $HCO_3^-(aq) + OH^-(aq) \longrightarrow$ _____

PREPARATION OF THE HALOGENS

The halogens normally occur in nature in the 1− oxidation state. Consequently, to prepare a halogen from a halide ion (X^-) the anion must be oxidized. This is possible by chemical means for all of the halogens but fluorine. Since fluorine is the most electronegative of all the elements, its resistance to oxidation should not be surprising. We can also see from the table of Standard Electrode Potentials that fluorine is the best oxidizing agent in the table. Both of these facts should lead us to expect that no chemical agent is capable of taking electrons from F^-. Therefore, F_2 is prepared by electrolysis. The procedure involves the electrolysis of HF dissolved in molten KF, with the anode reaction being

$$2\ F^-(l) \xrightarrow{\text{elect.}} F_2(g) + 2\ e^-$$

Chlorine. Chlorine is an excellent oxidizing agent, but there are a number of oxidizing agents that can be used to oxidize the Cl^- ion. The common ones are MnO_4^-, MnO_2, and $Cr_2O_7^{2-}$. These are used in the laboratory preparation of Cl_2. Accordingly, we have

$$2\ MnO_4^-(aq) + 16\ H^+(aq) + 10\ Cl^-(aq) \longrightarrow$$
$$2\ Mn^{2+}(aq) + 5\ Cl_2(g) + 8\ H_2O$$

$$4\ H^+(aq) + 2\ Cl^-(aq) + MnO_2(s) \longrightarrow$$
$$Mn^{2+}(aq) + 2\ H_2O + Cl_2(g)$$

$$14\ H^+(aq) + 6\ Cl^-(aq) + Cr_2O_7^{2-}(aq) \longrightarrow$$
$$2\ Cr^{3+}(aq) + 7\ H_2O + 3\ Cl_2(g)$$

Commercially, Cl_2 is now prepared by the electrolysis of molten NaCl or a brine (NaCl) solution.

Bromine. Bromine is produced commercially from Br^- ions present in either sea water or Arkansas brine wells. Note from the table of Standard Electrode Potentials that chlorine is a better oxidizing agent than bromine. Thus Cl_2 can be used to oxidize the Br^- ion.

$$Cl_2(g) + Br^-(aq) \longrightarrow Br_2(l) + 2\ Cl^-(aq)$$

The Br_2 formed by this method is rather impure. It is carried by an airstream into a Na_2CO_3 solution where the Br_2 disproportionates, giving

$$3\ Br_2(l) + 3\ CO_3^{2-}(aq) \longrightarrow 5\ Br^-(aq) + BrO_3^-(aq) + 3\ CO_2(g)$$

_____ **EXERCISE** _____

9.20 Br_2 is oxidized to _____ and is also reduced to _____ .

=================

This system is then acidified and the Br_2 is reobtained in pure form.

$$BrO_3^-(aq) + 5\ Br^-(aq) + 6\ H^+(aq) \longrightarrow 3\ Br_2(l) + 3\ H_2O$$

_____ **EXERCISE** _____

9.21 In the above reaction _____ is the oxidizing agent and _____ is the reducing agent.

=================

Iodine. Iodine is obtained from two different sources. It occurs in seaweed as the iodide ion (I^-) and in Chilean $NaNO_3$ ores as the iodate ion (IO_3^-). In the first case an oxidizing agent is needed to generate I_2, and in the second case a reducing agent is needed.

In the commercial preparation of I_2 from the iodide ion (I^-), Cl_2 is again used as the oxidizing agent. It may have occurred to you that both F_2 and Cl_2 can be used to oxidize the Br^- ion and that F_2, Cl_2, and Br_2 can be used to oxidize the I^- ion. The choice of Cl_2 is based on economics. It should also be understood that the oxidizing agents used for the laboratory preparation of Cl_2—namely, MnO_4^-, MnO_2, and $Cr_2O_7^{2-}$—can also be used for the preparation of Br_2 and I_2.

_____ **EXERCISE** _____

9.22 Let's see how you can do with these. Just give the products.

a. $I^-(aq) + Cl_2(g) \longrightarrow$ _____

b. $I^-(aq) + MnO_2(s) + H^+(aq) \longrightarrow$ _____

c. $MnO_4^-(aq) + H^+(aq) + Br^-(aq) \longrightarrow$ _____

d. $Cr_2O_7^{2-}(aq) + H^+(aq) + I^-(aq) \longrightarrow$ _____

=================

As was pointed out earlier, when an iodate ore is used, a reducing agent is necessary to obtain elemental iodine. For this reaction the hydrogen sulfite ion is used as the reducing agent:

$$2\ IO_3{}^-(aq) + 5\ HSO_3{}^-(aq) \longrightarrow I_2(s) + 5\ SO_4{}^{2-}(aq) + 3\ H^+(aq) + H_2O$$

—————— **EXERCISE** ——————

9.23   Now try these. Again, just give the products.

a. $Br^-(aq) + F_2(g) \longrightarrow$ _____

b. $Cl^-(aq) + MnO_2(s) + H^+(aq) \longrightarrow$ _____

c. $F^-(aq) + Cl_2(g) \longrightarrow$ _____

d. $I^-(aq) + MnO_4{}^-(aq) + H^+(aq) \longrightarrow$ _____

e. $BrO_3{}^-(aq) + Br^-(aq) + H^+(aq) \longrightarrow$ _____

SULFUR DIOXIDE AS A POLLUTANT

There are a number of pollutants found in the atmosphere today, with the most common being CO_2, CO, hydrocarbons, NO, O_3, particulates, and SO_2. Although any one of these can be particularly harmful, we will consider only SO_2 here. It is interesting that SO_2 occurs naturally from such diverse sources as volcanic gases, the oceans, and bacterial action. But by far the largest portion of SO_2 in the atmosphere comes from human activities. The burning of oil and coal and the smelting of sulfide ores are the major contributors.

The effects of atmospheric SO_2 can vary from the annoyance of burning eyes to severe respiratory problems. But we can't write simple chemical reactions to describe these effects. However, there are some important reactions of SO_2 that we can show.

To a large extent atmospheric SO_2 results from the burning of oil and coal in power generation. Sulfur in the oil and coal is oxidized, giving

$$S_8(s) + 8\ O_2(g) \longrightarrow 8\ SO_2(g)$$

A further source of SO_2 is illustrated by the roasting of the zinc ore sphalerite:

$$2\ ZnS(s) + 3\ O_2(g) \longrightarrow 2\ SO_2(g) + 2\ ZnO(s)$$

The problem is compounded by the further oxidation of the SO_2 to SO_3, which can occur by a number of pathways. It can result from exposure to sunlight (photochemical), and it is found that dust particles can catalyze the reaction. Thus, we find

$$2\ SO_2(g) + O_2(g) \longrightarrow 2\ SO_3(g)$$

And, of course, the SO_3 can react with H_2O, giving

$$SO_3(g) + H_2O \longrightarrow H_2SO_4(aq)$$

An unfortunate and insidious problem that we all face from the excessive production of SO_2 is the destruction of many of our great historical monuments. A building such as the Parthenon in Athens or a marble statue is composed predominantly of $CaCO_3$. By now we should well know that this reacts with an acid (or an acid anhydride). The reactions of concern are

$$H_2SO_4(aq) + CaCO_3(s) \longrightarrow CaSO_4(s) + CO_2(g) + H_2O(l)$$

$$SO_2(g) + CaCO_3(s) \longrightarrow CaSO_3(s) + CO_2(g)$$

and $$SO_3(g) + CaCO_3(s) \longrightarrow CaSO_4(s) + CO_2(g)$$

In each case, the monument will be damaged—a great loss to mankind.

It is possible to remove the SO_2 from the stack gases. This procedure involves scrubbing with lime. The basic reactions are familiar to us by now:

$$CaCO_3(s) \xrightarrow{\Delta} CaO(s) + CO_2(g)$$

$$CaO(s) + SO_2(g) \xrightarrow{\Delta} CaSO_3(s)$$

This procedure will, in fact, work, but it creates an excessively large amount of solid waste.

Another problem associated with SO_2 production is acid rain. It has been found that in some regions of the world the rain has become excessively acidic. This can be primarily attributed to the presence in the atmosphere of SO_2 and SO_3 which react with H_2O to give the corresponding acids. Nitric acid (HNO_3) also makes some contribution. Nitrogen is oxidized at high temperatures (about 1100°C) to its various oxides. These compounds are often by-products of the internal combustion engine. If, for instance, NO is formed, it can be further oxidized to NO_2:

$$N_2(g) + O_2(g) \xrightarrow{\Delta} 2\ NO(g)$$

$$2\ NO(g) + O_2(g) \longrightarrow 2\ NO_2(g)$$

and, of course, we know that NO_2 reacts with H_2O (moisture in the atmosphere) to give HNO_3:

$$3\ NO_2(g) + H_2O \longrightarrow 2\ HNO_3(aq) + NO(g)$$

It is primarily the presence of H_2SO_4 and HNO_3 that produces the acid rain.

In addition to its contribution to the destruction of monuments, acid rain has depleted and, in fact, totally destroyed the fish populations in many lakes and has had a devastating effect on forests and crops in those same areas. It is likely that acid rain occurs quite generally throughout most of the industrialized world, but in some regions there is no longer a natural ability of the environment to counteract its effects. The problem is particularly acute in the Northeastern part of the United States, the Southeastern part of Canada, and parts of the Scandinavian countries.

VOLCANOES

Tourists exploring the slopes of Mt. St. Helens may be puzzled when they find significant amounts of free sulfur. They may logically imagine that the free sulfur is among the emissions from the volcano. Actually, there is a whole field of chemistry devoted to the study of volcanoes, but we will only look at this one phenomenon. The explanation for the sulfur can be found in the analysis of the redox reactions that occur during the eruption of the volcano. Hot H_2S emitted by the volcano is oxidized when it comes into contact with oxygen in the air:

$$2\ H_2S(g) + 3\ O_2(g) \xrightarrow{\Delta} 2\ SO_2(g) + 2\ H_2O$$

The SO_2 that is formed then reacts further with more H_2S emanating from the volcano and is reduced to free sulfur:

$$16\ H_2S(g) + 8\ SO_2(g) \longrightarrow 3\ S_8(s) + 16\ H_2O$$

THE OZONE LAYER

There are two points that need to be made with respect to the *ozone layer*[3] before we go ahead. First, it is extremely important to understand that the

---

[3] The ozone layer is a region of the upper atmosphere at an elevation of about 15 miles above the earth's surface.

ozone layer is essential to our survival. The sun emits light, some of which is in the ultraviolet region of the electromagnetic spectrum. Ultraviolet light is known to cause skin cancer, and, in fact, if it were free to pass through the upper atmosphere and reach the earth's surface, the present forms of plant and animal life could not exist. It is this layer of ozone in the stratosphere that limits the passage of ultraviolet light. Thus, any possible depletion of the ozone layer is of serious concern. Second, the ozone layer maintains a relatively constant concentration of O_3 by virtue of a cyclic process that is initiated by ultraviolet radiation (radiant energy is symbolized by hv). In a simplified form, the reactions are

$$O_2(g) + hv \longrightarrow O(g) + O(g)$$

$$O(g) + O_2(g) \longrightarrow O_3(g) + \text{heat}$$

$$O_3(g) + hv \longrightarrow O_2(g) + O(g)$$

$$O(g) + O(g) \longrightarrow O_2(g) + \text{heat}$$

Thus, the ultraviolet radiation from the sun is converted to heat energy. This process maintains the ozone layer and in so doing absorbs most of the ultraviolet radiation from the sun.

Now with these reactions in mind, we can consider two potential threats to the ozone layer that have received considerable attention in recent years. The first of these is the supersonic transport airplane (SST). There was a major effort to develop an SST in the United States as well as abroad. In the United States the program was dropped, but the French and the British collaborated to build the Concorde. One of the major arguments against the SST is that, because of the altitude at which it flies, it may seriously deplete the ozone layer. The reason is that nitric oxide (NO) is produced in the engines of an SST just as it is produced in an automobile engine. The NO can then react with the ozone as follows:

$$O_3(g) + NO(g) \longrightarrow NO_2(g) + O_2(g)$$

and $$NO_2(g) + O(g) \longrightarrow NO(g) + O_2(g)$$

The atomic oxygen is available as a consequence of the cyclic process between O_2 and O_3 that we considered above. Note that in this series of reactions, the NO is regenerated and is, therefore, a catalyst for the reaction. The net reaction is then

$$O_3(g) + O(g) \longrightarrow 2 O_2(g)$$

thereby decreasing the amount of ozone.

A second potential threat to the ozone layer is found in the chlorofluorocarbons typified by CF_2Cl_2 and $CFCl_3$. As a group, these compounds are referred to as *freons*. They are used as refrigerants and as aerosol propellants in such things as deodorants, shaving creams, and hair dressings. Their use as propellants has now been curtailed in the United States, but they are still used in other parts of the world.

The problem with the freons was quite unexpected. They are very unreactive compounds. Consequently, they stay around for a long time. This permits them to finally diffuse into the stratosphere where they can react by a photochemical process to form atomic chlorine, for instance,

$$CF_2Cl_2(g) + h\nu \longrightarrow CF_2Cl(g) + Cl(g)$$

The atomic chlorine can then react with the ozone, giving

$$Cl(g) + O_3(g) \longrightarrow ClO(g) + O_2(g)$$

and

$$ClO(g) + O(g) \longrightarrow Cl(g) + O_2(g)$$

Note that as with the reaction between NO and O_3 above, the Cl is regenerated and is, therefore, a catalyst. And as with the NO reaction, the net reaction here is also

$$O_3(g) + O(g) \longrightarrow 2\ O_2(g)$$

with a corresponding decrease in the amount of ozone.

The type of reaction that we see here is different from those with which we have previously been concerned. These are known as *free radical* reactions and are quite common in photochemical processes. It is outside the intent of this book to teach you to predict the products of free radical reactions. Consequently, we won't pursue these reactions any further. But we thought you might be interested in seeing some chemistry that is important to all people whether they are taking a chemistry course or not.

CAVES

An interesting reaction of $CaCO_3$, as well as other carbonates, is

$$CaCO_3(s) + H_2O + CO_2(g) \rightleftharpoons Ca^{2+}(aq) + 2\ HCO_3^-(aq)$$

Since H_2O normally contains some dissolved CO_2, we see here a chemical basis for the dissolving of limestone. If a limestone deposit is below the earth's surface and is surrounded by insoluble rocks, a cave may form by virtue of this reaction. In fact, two of the most famous caves in this country, Mammoth cave in Kentucky and Carlsbad Caverns in New Mexico, are both limestone caves.

Certainly, one of the most interesting characteristics of many caves is the large icicle-like structures that hang from the roof (stalactites) and rise up from the floor (stalagmites), sometimes meeting to form a column. These projections result from the reversal of the above reaction. Water that contains dissolved hydrogen carbonates leaks through cracks in the roof of the cave. The HCO_3^- ion breaks down because of the evaporation of H_2O and because of the escape of CO_2 resulting from a decrease in its partial pressure. In either case, the result is the formation of the metal carbonate, which deposits on the surface of the stalactite as the saturated solution runs down the icicle-like structure. If any of the solution reaches the tip and drips to the floor, it creates a stalagmite.

SOAPS AND DETERGENTS

The last topic we will discuss in this chapter is that of soaps and synthetic detergents. We have already mentioned these in our previous discussion of hard water, but we didn't mention their syntheses or structures. Soaps have been known for many centuries—long before their chemistry was understood. In the frontier days, fat was obtained from a bear or a pig by boiling the fat meat. Wood ashes were then leached with water, and the two, the fat and the solution from the ashes, were heated together. From this, soap was obtained. Actually soap is a salt that can be made by an acid-base reaction. The acid is a long- and straight-chained organic acid typified by stearic acid, $CH_3-(CH_2)_{16}-COOH$. Such acids are called *fatty* acids and are obtained by the hydrolysis of fats. In the example we used, the base obtained from the leaching of the ashes is KOH. Thus the soap-making reaction is, in effect,

$$CH_3-(CH_2)_{16}-COOH + KOH \longrightarrow CH_3-(CH_2)_{16}-COO^-K^+ + H_2O$$

The more conventional bar soaps today are made with NaOH rather than KOH, giving the sodium salt of the fatty acid. Note that one end of the soap is a long-chain organic group. This end is oil soluble. The other end, $-COO^-Na^+$, is water soluble. These divergent properties provide the key to the functioning of a soap, which can, in effect, cause oils and water to mix.

A problem that commonly arises with soaps is that many metal ions that may be dissolved in water will form insoluble salts with the soap; for example,

$$2\ CH_3(CH_2)_{16}COO^-Na^+(aq) + Ca^{2+}(aq) \longrightarrow$$

$$[CH_3(CH_2)_{16}COO]_2Ca(s) + 2\ Na^+(aq)$$

Here the precipitation of the calcium salt removes the soap from the solution. This, of course, is the behavior we saw earlier in our discussion of hard water.

The present answer to the problem of hard water is the use of a synthetic detergent. A synthetic detergent mimics the oil-soluble and water-soluble ends of a soap. An example might be[4]

$$CH_3-(CH_2)_{11}-\hexagon-SO_3^-Na^+$$

The Ca^{2+}, Mg^{2+}, Fe^{2+}, etc., salts of these compounds are water soluble, and consequently the precipitation common to soaps is not a problem.

A difficulty encountered with the first generation of synthetic detergents was that they were not biodegradable. Consequently, it was common to see sudsy rivers and streams. The organic ends of these detergents were branched, and apparently bacteria know the difference and don't particularly care for such structures in their diets. The problem was readily solved by making straight-chained organic components such as occur in soaps and in our illustration above of a synthetic detergent.

Detergents also contain a *builder*, and this offers another difficulty. Generally, polyphosphates such as $Na_5P_3O_{10}$ are used. The problem here is that phosphates are plant nutrients, and dumping them into our waterways causes rapid growth of plant life in lakes and ponds. Although it is not a settled issue, it is possible that this growth speeds up the eutrophication[5] of our lakes. The common alternatives to the phosphate builders are washing soda ($Na_2CO_3 \cdot 10\ H_2O$), and borax ($Na_2B_4O_7 \cdot 10\ H_2O$), but these have the disadvantage of being highly basic. So we find that what appears at first to be a solution to a long-standing problem ends up creating new and perhaps equally serious problems. But at the same time we see another example of the challenge chemistry offers in everyday

---

[4] The hexagon stands for C_6H_4, a special structure in organic chemistry that need not concern you at this time.
[5] Eutrophication is an environmental condition in which the increased nutrients in a body of water form an environment conducive to plant life rather than animal life.

life. There will always be problems, but we can hope that there will also always be answers.

————— EXERCISES —————

9.24   As a final test for this chapter, let's see how you can do with these. Don't bother to balance the equations.

a. $CaF_2(s) + H_2SO_4(l) \xrightarrow{\Delta}$ _____

b. $NO_2(g) + H_2O \longrightarrow$ _____

c. $SO_3(g) + H_2O \longrightarrow$ _____

d. $AgCl(s) + CN^-(aq) \longrightarrow$ _____

e. $CuCO_3(s) \xrightarrow{\Delta}$ _____

f. $NaNO_3(s) + H_2SO_4(l) \xrightarrow{\Delta}$

g. $CuO(s) + C(s) \xrightarrow{\Delta}$ _____

h. $Ca(HCO_3)_2(s) \xrightarrow{\Delta}$ _____

i. $IO_3^-(aq) + HSO_3^-(aq) \longrightarrow$ _____

j. $Cl_2(g) + Br^-(aq) \longrightarrow$ _____

k. $PbO(s) + CO(g) \xrightarrow{\Delta}$ _____

l. $MnO_2(s) + H^+(aq) + Cl^-(aq) \longrightarrow$ _____

9.25   If you think you need more practice, do these.

a. $H_2S(g) + SO_2(g) \xrightarrow{\Delta}$ _____

b. $NaHCO_3(s) + HOAc(aq) \longrightarrow$ _____

c. $CaO(s) + H_2O \longrightarrow$ _____

d. $CaO(s) + SiO_2(s) \xrightarrow{\Delta}$ _____

e. $ZnS(s) + O_2(g) \xrightarrow{\Delta}$ _____

f. $CaCO_3(s) \xrightarrow{\Delta}$ _____

g. $CO_2(g) + H_2O \longrightarrow$ _____

h. $H_2S(g) + O_2(g) \xrightarrow{\Delta}$ _____

i. $I^-(aq) + MnO_2(s) + H^+(aq) \longrightarrow$ _____

j. $Na_2O(s) + SO_3(g) \longrightarrow$ _____

k. $SnO_2(s) + C(s) \xrightarrow{\Delta}$ _____

l. $Ag(CN)_2{}^-(aq) + Zn(s) \longrightarrow$ _____

m. $H^+(aq) + Cl^-(aq) + Cr_2O_7{}^{2-}(aq) \longrightarrow$ _____

Congratulations! You have successfully completed a course in reaction chemistry. This knowledge should enable you to have a better appreciation of basic chemistry. In the short run, you should more fully enjoy any future chemistry courses that you might take. In the long run, your completion of this course should allow you to have a better understanding of your physical world.

ANSWERS TO EXERCISES
CHAPTER 1

1.1. **a.** p **b.** s **c.** d **d.** p **e.** p **f.** f **g.** d

1.2. **a.** representative **b.** transition **c.** representative
d. transition **e.** noble gas
f. inner transition, lanthanide, rare earth
g. inner transition, actinide

1.3. **a.** Rb, Mg, Br, Se **b.** Sc, Mo **c.** Am **d.** Xe **e.** Lu, Pm
f. Lu, Pm, Am **g.** Lu, Pm

1.4. **a.** 3rd, s **b.** 9, 5, 4 **c.** $1s^2 2s^2 2p^6 3s^2 3p^6$ **d.** p
e. $1s^2 2s^2 2p^6 3s^2 3p^5$

1.5. **a.** $1s^2 2s^2 2p^6 3s^2 3p^2$ **b.** $1s^2 2s^2 2p^6 3s^2 3p^4$
c. $1s^2 2s^2 2p^6 3s^2 3p^6 4s^2 3d^5$ **d.** $1s^2 2s^2 2p^6 3s^2 3p^6 4s^2 3d^{10} 4p^2$
e. $1s^2 2s^2 2p^6 3s^2 3p^6 4s^2 3d^{10} 4p^6 5s^2$

1.6. Using shorthand notation,
a. $[Ne]3s^2$ **b.** $[Kr]5s^2 4d^{10} 5p^5$ **c.** $[Ar]4s^2 3d^{10}$
d. $[Xe]6s^2 4f^{14} 5d^{10} 6p^1$ **e.** $[Xe]6s^2 5d^1 4f^{14}$ **f.** $[Ar]$ **g.** $[Ar]$
h. $[Ar]$ **i.** $[Ar]$ **j.** $[Ar]$

1.7. **a.** $ns^2 np^2$ **b.** $ns^2 np^5$ **c.** ns^2 **d.** $ns^2(n-1)d^1$ **e.** $ns^2 np^6$

1.8. Using shorthand notation,
a. $[Ar]4s^2 3d^{10}$, $[Kr]5s^2 4d^{10}$, $[Xe]6s^2 4f^{14} 5d^{10}$ **b.** IIB
c. $ns^2(n-1)d^{10}$

1.9. **a.** $\underset{ns}{\uparrow\downarrow}$ $\underset{np}{\uparrow\ \uparrow\ —}$ **b.** $\underset{ns}{\uparrow\downarrow}$ $\underset{np}{\uparrow\ \uparrow\ \uparrow}$ **c.** $\underset{ns}{\uparrow\downarrow}$ $\underset{np}{\uparrow\downarrow\ \uparrow\downarrow\ \uparrow}$

d. $\underset{ns}{\uparrow\downarrow}$ $\underset{(n-1)d}{\uparrow\downarrow\ \uparrow\downarrow\ \uparrow\downarrow\ \uparrow\downarrow}$ $\underset{np}{—\ —\ —}$ **e.** $\underset{ns}{\uparrow\downarrow}$ $\underset{np}{\uparrow\downarrow\ \uparrow\downarrow\ \uparrow\downarrow}$

1.10. **a.** Cl **b.** Cl **c.** N **d.** O **e.** F
1.11. **a.** C **b.** I **c.** Na **d.** I **e.** Pb
1.12. **a.** VIA, $ns^2 np^4$, $2-$ **b.** IA, ns^1, $1+$ **c.** VIIA, $ns^2 np^5$, $1-$

1.13. a. $1-$ **b.** $2-$ **c.** $3-$ **d.** $1+$ **e.** $1-$ **f.** $2-$
  g. $2+$ **h.** $1-$ **i.** $1+$ **j.** $2-$
1.14. a. $MgCl_2$ **b.** Na_2S **c.** CaF_2 **d.** SrI_2 **e.** K_2O
  f. Rb_2Se **g.** Ba_3P_2 **h.** CaS **i.** Na_3N **j.** Cs_2Te

1.15.   Valence   Oxidation state

	Valence	Oxidation state
a.	2	$2-$
b.	1	$1-$
c.	1	$1+$
d.	4	$4+$
e.	3	$3-$

1.16. a. $2-$ **b.** $1-$ **c.** $1+$ **d.** $1+$ **e.** $2-$

1.17. a. $\underset{ns}{\uparrow\downarrow}$ **b.** $\underset{ns}{\uparrow\downarrow}\ \underset{np}{\uparrow\downarrow\ \uparrow\ \uparrow}$ **c.** $\underset{ns}{\uparrow\downarrow}\ \underset{np}{\uparrow\downarrow\ \uparrow\downarrow\ \uparrow}$

d. $\underset{ns}{\uparrow\downarrow}\ \underset{np}{\uparrow\ \_\ \_}$ **e.** $\underset{ns}{\uparrow}\ \underset{np}{\_\ \_\ \_}$ **f.** $\underset{ns}{\uparrow\downarrow}\ \underset{np}{\uparrow\ \_\ \_}$

g. $\underset{ns}{\uparrow\downarrow}\ \underset{np}{\uparrow\ \uparrow\ \uparrow}$ **h.** $\underset{ns}{\uparrow\downarrow}\ \underset{np}{\uparrow\ \uparrow\ \_}$ **i.** $\underset{ns}{\uparrow\downarrow}\ \underset{np}{\_\ \_\ \_}$

j. $\underset{ns}{\uparrow\downarrow}\ \underset{np}{\uparrow\downarrow\ \uparrow\ \uparrow}$

1.18. a. $2+$ **b.** $2-, 4+, 6+$ **c.** $1-, 5+, 7+$ **d.** $3+, 1+$
  e. $1+$ **f.** $3+, 1+$ **g.** $5+, 3+, 3-$ **h.** $4+, 2+, 4-$
  i. $2+$ **j.** $6+, 4+, 2-$

Note: You are merely predicting possible oxidation states. In part g, Bi, a metal, does not actually have a negative oxidation state. A similar situation exists for Ge (part h).

1.19. a. Al_2S_3, Al_2S **b.** NF_3, NF_5 **c.** GeO, GeO_2
  d. $TeCl_4$, $TeCl_6$ **e.** $InCl$, $InCl_3$ **f.** $SnSe$, $SnSe_2$
  g. PbO, PbO_2 **h.** Li_3N **i.** Tl_2O, Tl_2O_3 **j.** IF_7, IF_5

Note: Al_2S (part a) does not exist. NF_5 (part b) also does not exist. It should be noted here that orbital theory predicts the nonexistence of NF_5, as there are no $2d$ orbitals. Thus N can only have eight valence electrons and therefore form only four electron pair bonds.

1.20. a. ZnO **b.** $NiCl_2$ **c.** Nb_2S_5 **d.** Nd_2Se_3 **e.** CuI_2, CuI
  f. Sc_2O_3 **g.** TmN **h.** FeF_2, FeF_3 **i.** ZrS_2 **j.** MoO_3

1.21.   Structure   Valence   Oxidation state

a. $\begin{array}{c} H \\ | \\ H-C-H \\ | \\ H \end{array}$ 4   4 −

b. $\begin{array}{c} H \quad H \\ | \quad | \\ H-C-C-H \\ | \quad | \\ H \quad H \end{array}$ 4   3 −

c. $\begin{array}{c} H \quad H \quad H \\ | \quad | \quad | \\ H-C-C-C-H \\ | \quad | \quad | \\ H \quad H \quad H \end{array}$ 4 $2\frac{2}{3}$ −

d. Cl−Hg−Hg−Cl   2   1 +

e. $\begin{array}{c} H \qquad H \\ \searrow \qquad \nearrow \\ N-N \\ \nearrow \qquad \searrow \\ H \qquad H \end{array}$ 3   2 −

Note: A very important principle may be learned here. Valences and oxidation states are *not* identical. Valence stands for the number of bonds an atom forms. Oxidation states are based upon a set of rules. The results of part c point out that the oxidation state of an element need not be a whole number.

1.22. a. $NaCl$ **b.** CO, CO_2 **c.** CaO **d.** $GeCl_4$, $GeCl_2$
e. TeO_3, TeO_2 **f.** Li_3N **g.** PbF_2, PbF_4 **h.** InI, InI_3
i. Sb_2O_3, Sb_2O_5 **j.** $FeCl_2$, $FeCl_3$ **k.** Tb_2S_3 **l.** NiF_2
m. Cu_2O, CuO **n.** $LuCl_3$ **o.** TiF_3, TiF_4 **p.** ZnO

CHAPTER 2

2.1. **a** $3+$ **b.** $3+$ **c.** $5+$ **d.** $2+$ **e.** $3+$ **f.** $3+$ **g.** $3+$
h. $6+$ **i.** $2+$ **j.** $1+$

2.2. **a.** $Ba + Br_2 \longrightarrow BaBr_2$ **b.** $2 K + I_2 \longrightarrow 2 KI$
c. $Ca + Cl_2 \longrightarrow CaCl_2$ **d.** $16 Gd + 3 S_8 \longrightarrow 8 Gd_2S_3$
e. $2 Cs + Br_2 \longrightarrow 2 CsBr$ **f.** $2 Sc + 3 F_2 \longrightarrow 2 ScF_3$
g. $2 Mg + O_2 \longrightarrow 2 MgO$ **h.** $2 La + 3 F_2 \longrightarrow 2 LaF_3$
i. $2 Zn + O_2 \longrightarrow 2 ZnO$ **j.** $2 Dy + 3 Br_2 \longrightarrow 2 DyBr_3$

2.3. **a.** peroxide **b.** oxide **c.** peroxide **d.** peroxide
e. oxide **f.** oxide **g.** superoxide **h.** oxide
i. superoxide **j.** oxide

2.4. **a.** $2 Ca + O_2 \longrightarrow 2 CaO$ **b.** $Ba + O_2 \longrightarrow BaO_2$
c. $2 Na + O_2 \longrightarrow Na_2O_2$ **d.** $2 Mg + O_2 \longrightarrow 2 MgO$
e. $4 Li + O_2 \longrightarrow 2 Li_2O$ **f.** $K + O_2 \longrightarrow KO_2$
g. $Rb + O_2 \longrightarrow RbO_2$ **h.** $2 Sr + O_2 \longrightarrow 2 SrO$
i. $2 H_2 + O_2 \longrightarrow 2 H_2O$ **j.** $Cs + O_2 \longrightarrow CsO_2$

2.5. **a.** CrO_3, $\underline{Cr_2O_3}$ **b.** $\underline{Al_2O_3}$, Al_2O **c.** $\underline{Bi_2O_3}$, Bi_2O_5
d. $\underline{SnO}$, SnO_2 **e.** $\underline{Tl_2O}$, Tl_2O_3 **f.** WO_3

2.6. **a.** $\underline{CO_2}$, CO **b.** Sb_2O_5, Sb_2O_3 **c.** GeO, GeO_2
d. $\underline{SeO_2}$, SeO_3 **e.** P_4O_6, $\underline{P_4O_{10}}$ **f.** SO_2, $\underline{SO_3}$
g. $\underline{TeO_2}$, TeO_3 **h.** As_2O_3, As_2O_5

Note: Sb, Ge, Se, and As (parts b, c, d, and h) are located in the middle of their respective families; thus one cannot tell, by periodic positioning alone, which compound is the more stable.

2.7. **a.** $1-$ **b.** $1-$ **c.** $1+$ **d.** $1-$ **e.** $1+$ **f.** $1+$
g. $1+$ **h.** $1+$ **i.** $1-$ **j.** $1-$

2.8. **a.** LiH **b.** BaH_2 **c.** RbH **d.** CaH_2 **e.** NaH **f.** CsH

2.9. **a.** NR **b.** HI **c.** HCl **d.** NR **e.** HF **f.** H_2O
g. NR **h.** HBr **i.** NR **j.** NH_3

2.10. **a.** $3+, 1+$ **b.** $1+, 3+$ **c.** $2+$ **d.** $1+$ **e.** $2+, 4+$
f. $3+, 6+$ **g.** $2+$ **h.** $6+$ **i.** $3+, 5+$ **j.** $2+$

2.11. a. $AlCl_3$ **b.** $TlBr$ **c.** NiF_2 **d.** NaI **e.** $SnBr_2$ **f.** CrI_3
      g. CdF_2 **h.** WCl_6 **i.** BiI_3 **j.** $SrBr_2$

2.12. a. HBr **b.** $SbCl_5$ **c.** PI_3 **d.** $AsBr_3$ **e.** PCl_5
      f. SbI_3 (no pentaiodides exist)   **g.** $AsCl_3$ **h.** HI
      i. AsF_5 **j.** $BiBr_3$

2.13. SCl_4, TeF_6, $SeCl_4$, SF_6

2.14. a. ICl, IBr, BrF, $BrCl$, ClF **b.** ClF_3, BrF_3, ICl_3
      c. IF_5, BrF_5 **d.** IF_7

2.15. a. SnS, (SnS_2) **b.** Sb_2S_3 **c.** WCl_6 **d.** $CrCl_3$ **e.** Na_2Se
      f. Bi_2S_3 **g.** MoO_3 **h.** SiS_2 **i.** K_2S **j.** $CaSe$ **k.** BaS
      l. CS_2 **m.** Rb_2Te **n.** Li_3N

REVIEW I

I.1.

	Complete	Shorthand	Family	Orbital
b.	$1s^2 2s^2 2p^6 3s^2 3p^2$	$[Ne]3s^2 3p^2$	$ns^2 np^2$	$\underset{ns}{\uparrow\downarrow}\ \underset{np}{\uparrow\ \uparrow\ \underline{}}$
c.	$1s^2 2s^2 2p^6 3s^2 3p^3$	$[Ne]3s^2 3p^3$	$ns^2 np^3$	$\underset{ns}{\uparrow\downarrow}\ \underset{np}{\uparrow\ \uparrow\ \uparrow}$
d.	$1s^2 2s^2 2p^6 3s^2 3p^6$	$[Ar]$	$ns^2 np^6$	$\underset{ns}{\uparrow\downarrow}\ \underset{np}{\uparrow\downarrow\ \uparrow\downarrow\ \uparrow\downarrow}$
e.	$1s^2 2s^2 2p^6 3s^2 3p^6$ $4s^2 3d^3$	$[Ar]4s^2 3d^3$	$ns^2(n-1)d^3$	$\underset{ns}{\uparrow\downarrow}\ \underset{(n-1)d}{\uparrow\ \uparrow\ \uparrow\ \underline{}\ \underline{}}$
f.	$1s^2 2s^2 2p^6 3s^2 3p^6$ $4s^2 3d^{10} 4p^3$	$[Ar]4s^2 3d^{10} 4p^3$	$ns^2 np^3$	$\underset{ns}{\uparrow\downarrow}\ \underset{np}{\uparrow\ \uparrow\ \uparrow}$
g.	$1s^2 2s^2 2p^6 3s^2 3p^6 3d^{10}$	$[Ar]3d^{10}$		$\underset{ns}{\uparrow\downarrow}\ \underset{np}{\uparrow\downarrow\ \uparrow\downarrow\ \uparrow\downarrow}$ $\underset{nd}{\uparrow\downarrow\ \uparrow\downarrow\ \uparrow\downarrow\ \uparrow\downarrow\ \uparrow\downarrow}$
h.	$1s^2 2s^2 2p^6 3s^2 3p^6$	$[Ar]$		$\underset{ns}{\uparrow\downarrow}\ \underset{np}{\uparrow\downarrow\ \uparrow\downarrow\ \uparrow\downarrow}$
i.	$1s^2 2s^2 2p^6 3s^2 3p^6$ $4s^2 3d^{10} 4p^6$	$[Kr]$		$\underset{ns}{\uparrow\downarrow}\ \underset{np}{\uparrow\downarrow\ \uparrow\downarrow\ \uparrow\downarrow}$
j.	$1s^2 2s^2 2p^6 3s^2 3p^6$	$[Ar]$		$\underset{ns}{\uparrow\downarrow}\ \underset{np}{\uparrow\downarrow\ \uparrow\downarrow\ \uparrow\downarrow}$

Note: The Ga^{3+} ion has an $[Ar]$ configuration plus completely filled d orbitals. Thus the $3d^{10}$ is included to differentiate it from Sc^{3+} and Cl^- (parts h and j), which only have 18 electrons.

I.2. b. Cr, W   **c.** all lanthanides   **d.** C, Si, Sn, Pb
     e. F, Cl, Br, At   **f.** Ti, Hf   **g.** He, Ne, Ar, Xe, Rn
     h. B, Al, Ga, In   **i.** Zn, Cd   **j.** O, S, Se, Po

Note: Pr is a lanthanide, which means that its family members are in the row beginning with Ce and ending with Lu.

I.3. b. transition   **c.** noble gas
     d. inner transition, or lanthanide, or rare earth   **e.** representative
     f. inner transition, or actinide, or second rare earth   **g.** transition
     h. representative   **i.** transition   **j.** noble gas

I.4. a. 2+   **b.** 7+, 5+, 1−   **c.** 6+, 4+, 2−   **d.** 1+   **e.** 3+
     f. 3+, 1+   **g.** 3+   **h.** 3+, 1+   **i.** 2+   **j.** 3+, 1+   **k.** 0
     l. 5+, 3+, 3−   **m.** 2+   **n.** 4+, 2+   **o.** 6+, 4+, 2−

I.5. a. 3+   **b.** 2+   **c.** 3+, 2+   **d.** 1+   **e.** 2+
     f. 5+, 4+, 3+   **g.** 4+   **h.** 3+, 2+   **i.** 6+
     j. 7+, 4+, 3+, 2+

I.6. a. Cl   **b.** S   **c.** Cl   **d.** N   **e.** Na   **f.** Si   **g.** I   **h.** I
     i. S   **j.** Cl   **k.** H   **l.** H   **m.** C   **n.** H   **o.** H

I.7. a. Tl_2O, Tl_2O_3 **b.** CBr_4 **c.** Na_3N **d.** Al_2S_3
     e. BiN, Bi_3N_5 **f.** $CrCl_3$, $CrCl_6$ **g.** Pm_2S_3 **h.** PbF_2, PbF_4
     i. P_4O_{10}, P_4O_6 **j.** FeS, Fe_2S_3

Note: One might predict the existence of CBr_2. It turns out, however, that the only common carbon compound possessing a 2+ oxidation state is carbon monoxide, CO.

I.8. a. $InCl$, $InCl_3$ **b.** Sc_2O_3 **c.** Li_3N **d.** SO_2, SO_3
     e. $CrBr_3$, $CrBr_6$ **f.** $ZnCl_2$ **g.** Al_2O_3 **h.** NaF
     i. $PrCl_3$ **j.** GeO, GeO_2 **k.** TeF_4, TeF_6 **l.** SnI_2, SnI_4
     m. Tl_2O, Tl_2O_3 **n.** PBr_3, PBr_5 **o.** $SbCl_3$, $SbCl_5$ **p.** FeS, Fe_2S_3

I.9. a. CaO **b.** RbO_2 **c.** $AlCl_3$ **d.** Li_2O **e.** $GeCl_2$ or $GeCl_4$
     f. Tl_2O **g.** Na_2O_2 **h.** PbI_2 **i.** GaF_3 **j.** PbF_4
     k. CsO_2 **l.** $TlBr$ **m.** KH **n.** PCl_3 **o.** Al_2O_3 **p.** BaO_2

Note: Ge (part e) is in the middle of its family; thus one cannot determine which is the more likely product. Although it is true that Ga (part i) is also placed in the middle of its family, fluorine (see rule 4) takes an element to its highest oxidation state.

3.1. **a.** sodium fluoride   **b.** potassium iodide   **c.** magnesium oxide
     d. rubidium sulfide   **e.** barium bromide   **f.** potassium hydride
     g. silver chloride

3.2. **a.** rubidium superoxide   **b.** calcium peroxide
     c. cesium superoxide   **d.** barium peroxide

3.3. **a.** potassium oxide   **b.** sodium oxide   **c.** sodium peroxide
     d. potassium superoxide

3.4. **a.** gold(III) chloride   **b.** nickel(II) bromide   **c.** copper(I) iodide
     d. iron(III) bromide   **e.** chromium(III) sulfide

3.5. **a.** ferric oxide   **b.** stannous oxide   **c.** cupric bromide
     d. mercurous chloride

3.6. **a.** silicon tetrachloride   **b.** hydrogen chloride
     c. iodine heptafluoride   **d.** iodine tribromide
     e. dinitrogen trioxide   **f.** phosphorus pentachloride
     g. phosphorus trichloride

3.7. *Preferred name* listed first in parts b, c, and d.
     a. manganese(II) oxide
     b. manganese(III) oxide or dimanganese trioxide
     c. manganese(IV) oxide or manganese dioxide
     d. manganese(VII) oxide or dimanganese heptoxide

3.8. **a.** (ortho)phosphate   **b.** sulfite   **c.** nitrite
     d. dichromate   **e.** chlorite   **f.** sulfate
     g. permanganate   **h.** perchlorate   **i.** chromate
     j. nitrate   **k.** ammonium   **l.** chlorate
     m. carbonate   **n.** arsenite

3.9. **a.** NO_3^- **b.** SO_4^{2-} **c.** $Cr_2O_7^{2-}$ **d.** OH^- **e.** ClO_3^-
     f. AsO_2^- or AsO_3^{3-} **g.** NH_4^+ **h.** CrO_4^{2-} **i.** NO_2^-
     j. PO_4^{3-} **k.** ClO_4^- **l.** MnO_4^- **m.** SO_3^{2-} **n.** CO_3^{2-}

3.10. **a.** sodium arsenite   **b.** barium hydroxide
     c. zinc nitrate   **d.** iron(II) sulfite

  e. copper(II) carbonate **f.** calcium (ortho)phosphate
  g. silver sulfate **h.** mercury(II) chlorate
3.11. a. ammonium carbonate **b.** calcium hydrogen carbonate
  c. sodium dihydrogen (ortho)phosphate
3.12. a. rubidium perchlorate **b.** calcium (ortho)phosphate
  c. silver chromate **d.** sodium hypochlorite
  e. ammonium cyanide **f.** zinc iodate **g.** aluminum sulfate
  h. diantimony pentasulfide **i.** arsenic(III) sulfite
  j. tellurium(VI) oxide or tellurium trioxide

Note: In part j, Te may be considered to be an element whose properties are both metallic and nonmetallic. Thus both metallic and nonmetallic names have been given.

3.13. a. iron(III) hydroxide **b.** cesium hydrogen carbonate
  c. nickel(II) (ortho)phosphate **d.** mercury(II) chloride
  e. magnesium permanganate **f.** lead(IV) fluoride
  g. tin(II) dichromate **h.** dichlorine heptoxide
  i. cobalt(II) hydrogen (ortho)phosphate **j.** aluminum nitrite

CHAPTER 4

4.1. **a.** 4+ **b.** 3− **c.** 5+ **d.** 3+ **e.** 5+ **f.** 6+
     g. 7+ **h.** 5+ **i.** 6+ **j.** 3+ **k.** 0 **l.** 2+

4.2. **a.** 6+ **b.** 5+ **c.** 6+ **d.** 7+ **e.** 4+ **f.** 3+
     g. 3+ **h.** 1+ **i.** 7+ **j.** 6+ **k.** 6+ **l.** 1+

4.3. **a.** AsO_4^{3-} or AsO_3^{-} **b.** BrO_4^{-} **c.** TeO_3^{2-} **d.** SiO_3^{2-}
     e. MoO_4^{2-} **f.** IO_3^{-} **g.** SnO_3^{2-} **h.** AsO_3^{3-} or AsO_2^{-}
     i. SeO_4^{2-} **j.** PH_4^{+}

4.4. **b.** $HAsO_2$ AsO_2^{-} metaarsenite ion
           H_3AsO_3 AsO_3^{3-} orthoarsenite ion
     c. $HSbO_3$ SbO_3^{-} metaantimonate ion
           H_3SbO_4 SbO_4^{3-} mesoantimonate ion
           H_5SbO_5 SbO_5^{5-} orthoantimonate ion
     d. H_2SeO_3 SeO_3^{2-} metaselenite ion
           H_4SeO_4 SeO_4^{4-} orthoselenite ion
     e. $HAlO_2$ AlO_2^{-} metaaluminate ion
           H_3AlO_3 AlO_3^{3-} orthoaluminate ion
     f. H_2SnO_3 SnO_3^{2-} metastannate ion
           H_4SnO_4 SnO_4^{4-} orthostannate ion

4.5. HIO_3 IO_3^{-}
       H_3IO_4 IO_4^{3-}
       H_5IO_5 IO_5^{5-}

4.6. **a.** orthoborate   **b.** metaphosphate   **c.** metabromate
     d. orthoselenite   **e.** metaaluminate   **f.** orthoaluminate
     g. orthogermanate   **h.** metaarsenate

Note: All the oxyacids formed from the orthoanions have the form
$M(OH)_x$. Thus the true orthooxyacids would be H_3BO_3, H_4SeO_4, and
H_3AlO_3. Remember that orthophosphoric acid, H_3PO_4, is actually mis-
named. See p. 69.

4.7. b, h, i, l

4.8. b, d, j

4.9. **b.** $K^+ + ClO_3^-$ **c.** $Ba^{2+} + 2I^-$ **d.** $Cs^+ + Br^-$
   e. $2K^+ + Cr_2O_7^{2-}$ **f.** $2NH_4^+ + SO_4^{2-}$ **g.** $2Rb^+ + SO_3^{2-}$
   h. $2Al^{3+} + 3SO_4^{2-}$ **i.** $2Ga^{3+} + 3CO_3^{2-}$ **j.** $3Cs^+ + PO_4^{3-}$
   k. $K^+ + MnO_4^-$ **l.** $Rb^+ + NO_3^-$ **m.** $3NH_4^+ + PO_4^{3-}$
   n. $Ca^{2+} + 2I^-$ **o.** $NH_4^+ + Cl^-$ **p.** $Bi^{3+} + 3NO_3^-$

4.10. b, d, e, g, j, n, o

4.11. **b.** $SnCO_3(s) + 2Na^+ + 2Cl^-$ **c.** $2AgCl(s) + Ba^{2+} + 2NO_3^-$
   d. $BaSO_4(s) + 2Ag^+ + 2NO_3^-$ **e.** $Ag_3PO_4(s) + 3Na^+ + 3NO_3^-$
   f. NR   **g.** $SrCO_3(s) + 2NH_4^+ + 2Cl^-$
   h. $Sn_3(PO_4)_2(s) + 6Na^+ + 6ClO_3^-$

4.12. **a.** $Ag_2S(s) + 2Na^+ + 2ClO_4^-$ **b.** NR   **c.** NR   **d.** NR
   e. $FeS(s) + 2Rb^+ + 2ClO_3^-$ **f.** $PbCrO_4(s) + 2NH_4^+ + 2NO_3^-$
   g. NR   **h.** NR

II.1.   **a.** zinc chloride   **b.** strontium oxide   **c.** iron(II) iodide
      d. copper(I) sulfide   **e.** lead(IV) fluoride   **f.** dinitrogen oxide
      g. phosphorus trichloride   **h.** sulfur hexafluoride
      i. dinitrogen pentasulfide   **j.** disulfur dichloride

II.2.   **a.** sodium oxide   sodium peroxide
      b. potassium oxide   potassium superoxide
      c. cesium oxide   cesium superoxide
      d. barium oxide   barium peroxide

II.3.   **a.** ammonium iodide   **b.** sodium nitrate   **c.** copper(II) sulfate
      d. lead(II) carbonate   **e.** potassium nitrite
      f. cadmium orthophosphate   **g.** mercury(II) chlorate
      h. iron(II) sulfite

II.4.   **a.** dichromate   **b.** hypochlorite   **c.** perchlorate
      d. cyanide   **e.** hydroxide   **f.** chlorite

II.5.   **a.** ammonium nitrate   **b.** ammonium orthophosphate
      c. ammonium sulfate

II.6.   **a.** sulfate   **b.** hydrogen sulfate   **c.** orthophosphate
      d. hydrogen orthophosphate   **e.** carbonate
      f. hydrogen carbonate   **g.** dihydrogen orthophosphate
      h. dihydrogen orthophosphite

II.7.   **a.** potassium orthophosphate
      b. potassium hydrogen orthophosphate
      c. potassium dihydrogen orthophosphate

II.8.   **a.** 6+   **b.** 5+   **c.** 7+   **d.** 3+   **e.** 4+   **f.** 5+

II.9.   Formula of ion   Name of ion

      a.   _____   carbonate

      b. $SiO_3{}^{2-}$ silicate

c. $GeO_3{}^{2-}$ germanate

d. $SnO_3{}^{2-}$ _____

II.10.   Anion   Prefix   Name of acid

a. $SnO_3{}^{2-}$ _____   metastannic

b. $SnO_4{}^{4-}$ *ortho-*   _____

c. $AlO_2{}^-$ *meta-*   metaaluminic

d. $AlO_3{}^{3-}$ *ortho-*   orthoaluminic

e. $IO_3{}^-$ *meta-*   metaiodic

f. $IO_4{}^{3-}$ *meso-*   mesoiodic

g. $IO_5{}^{5-}$ *ortho-*   orthoiodic

II.11. a, c, d, e, f, j, k

II.12. **a.** $Na^+ + Cl^-$ **b.** $Mg^{2+} + 2 NO_3{}^-$ **c.** $Fe^{3+} + 3 Cl^-$
        d. $2 NH_4{}^+ + S^{2-}$ **e.** $3 K^+ + PO_4{}^{3-}$ **f.** $2 NH_4{}^+ + SO_4{}^{2-}$
        g. $K^+ + H_2PO_4{}^-$

II.13.
b. $ZnCl_2$ + K_2SO_4 $\longrightarrow$ $ZnSO_4(s) + 2 K^+ + 2 Cl^-$
        zinc chloride + potassium sulfate $\rightarrow$ zinc sulfate + potassium chloride

c. $3 AgNO_3$ + Na_3PO_4 $\longrightarrow$ $Ag_3PO_4(s)$ + $3 Na^+$ + $3 NO_3{}^-$
        silver nitrate +   sodium $\rightarrow$ silver   +   sodium nitrate
                            orthophosphate   orthophosphate

d. $(NH_4)_2SO_4$ + $Pb(NO_3)_2$ $\longrightarrow$ $PbSO_4(s) + 2 NH_4{}^+ + 2 NO_3{}^-$
        ammonium sulfate + lead(II) nitrate $\rightarrow$ lead(II) sulfate + ammonium nitrate

e. $Hg_2(NO_3)_2$ + $2 NaCl$ $\longrightarrow$ $Hg_2Cl_2(s)$ + $2 Na^+$ + $2 NO_3{}^-$
        mercury(I) nitrate + sodium chloride $\rightarrow$ mercury(I) chloride + sodium nitrate

f. $3 CaS$ + $2 FeCl_3$ $\longrightarrow$ $Fe_2S_3(s)$ + $3 Ca^{2+}$ + $6 Cl^-$
        calcium sulfide + iron(III) chloride $\rightarrow$ iron(III) sulfide + calcium chloride

g. Na_2CrO_4 + $2 AgNO_3$ $\longrightarrow$ $Ag_2CrO_4(s) + 2 Na^+ + 2 NO_3{}^-$
        sodium chromate + silver nitrate $\rightarrow$ silver chromate + sodium nitrate

h. $2 (NH_4)_3PO_4$ + $3 CuCl_2$ $\longrightarrow$ $Cu_3(PO_4)_2(s) + 6 NH_4{}^+ + 6 Cl^-$
        ammonium   +   copper(II) $\rightarrow$ copper(II)   + ammonium chloride
        orthophosphate   chloride   orthophosphate

ANSWERS TO EXERCISES
CHAPTER 5

5.1. **a.** *hydro-* **b.** *oxy-* **c.** *hydro-* **d.** *oxy-* **e.** *hydro-* **f.** *oxy-*
5.2. **a.** hydrobromic **b.** sulfurous **c.** nitrous **d.** hydroselenic
e. perchloric **f.** metaphosphoric **g.** chromic
h. hydrofluoric **i.** metaarsenic **j.** metaselenic
k. chlorous **l.** orthotelluric **m.** metaiodic
n. hydrocyanic **o.** orthophosphorous **p.** hypochlorous
5.3. **a.** $H_2 + Cl_2 \longrightarrow 2\ HCl$ **b.** $H_2 + F_2 \longrightarrow 2\ HF$
c. $H_2 + Br_2 \longrightarrow 2\ HBr$ **d.** $H_2 + I_2 \longrightarrow 2\ HI$
5.4. **a.** $7+; 7+$ **b.** $3+; 5+, 3+$ **c.** $6+; 6+, 4+$ **d.** $5+; 5+, 3+$
e. $5+; 5+, 3+$ **f.** $4+; 4+, 2+$ **g.** $6+; 6+, 4+$
h. $4+; 6+, 4+$ **i.** $7+; 7+, 5+$ **j.** $6+; 6+, 4+$
5.5. **a.** yes, $\overset{6+}{H_2SO_4}$ **b.** yes, $\overset{4+}{H_2CO_3}$ **c.** no **d.** no
e. yes, $\overset{7+}{HClO_4}$ **f.** yes, $\overset{3+}{H_3PO_3}$ **g.** no **h.** yes, $\overset{4+}{H_2SeO_3}$
i. yes, $\overset{5+}{HNO_3}$ **j.** no
5.6. **a.** H_2CO_3 **b.** HIO_3 **c.** H_2SO_3 **d.** H_2SeO_3 **e.** H_3PO_3
f. $HClO_4$ **g.** HNO_3 **h.** H_2SO_4 **i.** $H_3PO_4\ (HPO_3)$
j. $H_3AsO_3\ (HAsO_2)$
5.7. **a.** $Na_2SO_4 + HNO_3$ **b.** $Na_2SO_4 + HCN$ **c.** $BaSO_4 + HClO_2$
d. $Na_3PO_4 + HI$ **e.** $K_3PO_4 + HBr$ **f.** $Na_2SO_4 + HCl$
g. $BaSO_4 + HNO_3$ **h.** $K_2SO_4 + HF$ **i.** $BaSO_4 + HClO_4$
j. $Na_2SO_4 + HBr$
5.8. **a.** $HI + H_3PO_3$ **b.** $HBr + H_3PO_4$ **c.** $HCl + H_3PO_4$
d. $HCl + H_3PO_3$

5.9. $H_2 + Cl_2 \longrightarrow 2\ HCl$

$PCl_3 + 3\ H_2O \longrightarrow 3\ HCl + H_3PO_3$

$H_2SO_4 + 2\ NaCl \xrightarrow{\Delta} Na_2SO_4 + 2\ HCl$

$BaCl_2 + H_2SO_4 \xrightarrow{H_2O} 2\ HCl + BaSO_4(s)$

5.10. $H_2SO_4 + 2\,NaIO_3 \xrightarrow{\Delta} Na_2SO_4 + 2\,HIO_3$

$I_2O_5 + H_2O \longrightarrow 2\,HIO_3$

$Ba(IO_3)_2 + H_2SO_4 \xrightarrow{H_2O} 2\,HIO_3 + BaSO_4(s)$

5.11. a. $P_4O_{10} + 6\,H_2O \longrightarrow 4\,H_3PO_4$

b. $PCl_5 + 4\,H_2O \longrightarrow 5\,HCl + H_3PO_4$

5.12. a. $N_2O_3 + H_2O \longrightarrow 2\,HNO_2$

b. $H_2SO_4 + BaNO_2 \xrightarrow{H_2O} BaSO_4(s) + 2\,HNO_2$

5.13. a. superoxide, no   **b.** oxide, yes   **c.** peroxide, no
d. oxide, yes   **e.** oxide, no   **f.** superoxide, no   **g.** oxide, yes
h. peroxide, no   **i.** superoxide, no   **j.** oxide, yes

5.14. a. $Ba(OH)_2$ **b.** $La(OH)_3$ **c.** $ZnO + H_2O$ **d.** $NaOH$
e. $K_2O + H_2O$ **f.** $Al(OH)_3$ **g.** $AgOH$ **h.** $Ca(OH)_2$
i. $SrO + H_2O$ **j.** $CsOH$

5.15. b. $KCl + H_2O$ **c.** $CsI + H_2O$ **d.** $AgCl + H_2O$
e. $RbBr + H_2O$

5.16. b. $CaBr_2 + 2\,H_2O$ **c.** $K_2SO_4 + 2\,H_2O$
d. $Cs_3PO_4 + 3\,H_2O$ **e.** $BaSO_4 + 2\,H_2O$
f. $Al_2(SO_4)_3 + 6\,H_2O$ **g.** $Sr_3(PO_4)_2 + 6\,H_2O$

5.17. i, j

5.18.

	Base	Conjugate acid
a. $H_3O^+ + NO_2^-$	H_2O	H_3O^+
b. $H_3O^+ + CN^-$	H_2O	H_3O^+
c. $NH_4^+ + ClO_2^-$	NH_3	NH_4^+
d. $H_3O^+ + HS^-$	H_2O	H_3O^+
e. $NH_4^+ + CN^-$	NH_3	NH_4^+

5.19.

	Acid	Conjugate base
a. $HClO_2 + OH^-$	H_2O	OH^-
b. $H_3PO_3 + NH_2^-$	NH_3	NH_2^-
c. $HNO_2 + OH^-$	H_2O	OH^-
d. $H_3O^+ + HCO_3^-$	H_2CO_3	HCO_3^-
e. $H_2SO_3 + OH^-$	H_2O	OH^-
f. $NH_4^+ + NO_2^-$	HNO_2	NO_2^-
g. $H_3O^+ + H_2PO_3^-$	H_3PO_3	$H_2PO_3^-$
h. $H_3PO_3 + OH^-$	H_2O	OH^-
i. $NH_4^+ + CN^-$	HCN	CN^-
j. $HCN + NH_2^-$	NH_3	NH_2^-

CHAPTER 6

6.1. HCl, HBr, HI, H_2SO_4, HNO_3, $HClO_4$, $HClO_3$

6.2. **a.** KOH, weak   **b.** strong, HCN

6.3.

	Acid	Base
a.	HI	NaOH
b.	HF	CsOH
c.	HBr	$Ca(OH)_2$
d.	HBr	$NH_3(aq)$
e.	$HClO_2$	$Fe(OH)_2$
f.	H_3PO_4	NaOH
g.	H_2SO_4	$Al(OH)_3$

6.4. **a.** acidic **b.** basic **c.** neutral **d.** acidic **e.** neutral
    f. basic **g.** neutral **h.** acidic **i.** acidic **j.** basic
    h. neutral

Note: The salts in parts a, d, h, and i are formed from strong acids and weak bases, whereas the salts in parts b, f, and j are formed from strong bases and weak acids. The neutral salts, those in parts c, e, g, and k, are formed from strong acids and strong bases.

6.5. **a.** $KCl + H_2O$ **b.** $NaNO_3 + H_2O$ **c.** $CaI_2 + 2\,H_2O$
    d. $BaBr_2 + 2\,H_2O$

6.6. **a.** $K^+(aq) + Cl^-(aq)$ **b.** $Na^+(aq) + NO_3{}^-(aq)$
    c. $Ca^{2+}(aq) + 2\,I^-(aq)$ **d.** $Ba^{2+}(aq) + 2\,Br^-(aq)$
    e. $K^+(aq) + ClO_4{}^-(aq)$

6.7. **a.** neutral   **b.** basic   **c.** acidic

6.8. H_2O, F^-

6.9. a, c, h, n

6.10. weak, strong, hydration, hydrolysis

6.11. $Na^+(aq)$

Note: The (aq) is included here to illustrate the hydration of Na^+. This procedure will be followed in each problem of this type.

6.12. $HCN + OH^-$

6.13. a. $Na^+ + CN^-$ **b.** $Na^+(aq)$ **c.** $HCN + OH^-$

6.14. base, acid, base, hydration

6.15. KF $NaOAc$ KNO_2

6.16. $F^- + H_2O \rightleftharpoons HF + OH^-$
$OAc^- + H_2O \rightleftharpoons HOAc + OH^-$
$NO_2^- + H_2O \rightleftharpoons HNO_2 + OH^-$

6.17. a. Na^+ **b.** OAc^- **c.** Na^+
d. $OAc^- + H_2O \rightleftharpoons HOAc + OH^-$

6.18. a. HBr, $NH_3(aq)$, $NH_3(aq) + HBr(aq) \longrightarrow NH_4Br$ **b.** $NH_4^+ + Br^-$
c. $Br^-(aq)$ **d.** weak, hydrolysis, $NH_3 + H_3O^+$
e. acidic salt; it is formed from a strong acid and a weak base.

6.19. NH_4NO_3, $Cd(ClO_3)_2$, CuI_2

6.20. a. $NaCN + H_2O \longrightarrow Na^+ + CN^-$
$Na^+ + H_2O \longrightarrow NR$
$CN^- + H_2O \rightleftharpoons HCN + OH^-$

NaCN is a basic salt. The spectator ion is Na^+ (it undergoes
hydration).

b. $NH_4NO_3 + H_2O \longrightarrow NH_4^+ + NO_3^-$
$NO_3^- + H_2O \longrightarrow NR$
$NH_4^+ + H_2O \rightleftharpoons NH_3 + H_3O^+$

NH_4NO_3 is an acidic salt. NO_3^- is the spectator ion.

c. $RbClO_4 + H_2O \longrightarrow Rb^+ + ClO_4^-$
$Rb^+ + H_2O \longrightarrow NR$
$ClO_4^- + H_2O \longrightarrow NR$

$RbClO_4$ is a neutral salt. There is no hydrolysis; therefore both ions
are spectator ions.

CHAPTER 7

7.1.   **a.** a proton donor   **b.** a proton acceptor

7.2.   **a.** HCl   **b.** H_2O

7.3.   acid, base

7.4.   **a.** H_2O **b.** HCN   **c.** $PH_4^+ + I^-$ **d.** BCl_4^-

7.5.   **a.** $NaCl + H_2O$ **b.** $BaSO_4 + H_2O$ **c.** $CaSO_4$
      d. $CaCO_3 + H_2O$ **e.** $MgSO_3$ **f.** $Cs_3PO_4 + H_2O$
      g. K_2CO_3 **h.** $SrCO_3 + H_2O$

Note: A salt is formed in each reaction. H_2O is a product only when both hydrogen and oxygen are present in the reactants.

7.6.   **a.** AgCl   **b.** Cl^-

7.7.   **a.** $Ag(CN)_2^-$ **b.** $Cr(H_2O)_6^{3+}$ **c.** $Fe(H_2O)_6^{3+}$
      d. $Co(NH_3)_6^{3+}$ **e.** $Cu(NH_3)_4^{2+}$ **f.** $PtCl_4^{2-}$
      g. $Zn(CN)_4^{2-}$ **h.** FeF_6^{3-} **i.** $Ti(H_2O)_6^{3+}$ **j.** CdI_4^{2-}

Note: In parts a, f, g, h, and j, the charge on the complex ion is different from the charge on the central atom, because the ligand in each case is charged.

7.8.   cyano, chromate, hexacyanochromate(III) ion

7.9.   **a.** hexachloroplatinate(IV) ion   **b.** hexabromocobaltate(III) ion
      c. hexacyanocobaltate(III) ion

7.10.   **a.** dichloroargentate(I) ion   **b.** tetrahydroxoaluminate(III) ion
      c. tetracyanoaurate(III) ion   **d.** hexacyanoferrate(III) ion
      e. tetracyanonickelate(0) ion

7.11.   **a.** tetraaquacopper(II) ion   **b.** diamminesilver(I) ion
      c. hexaaquachromium(III) ion   **d.** hexaaquairon(II) ion
      e. tetraamminenickel(II) ion

7.12.   **a.** hexaaquachromium(III) ion   **b.** tetracarbonylnickel(0)
      c. tetrahydroxozincate(II) ion   **d.** hexachloroferrate(III) ion
      e. diamminesilver(I) chloride
      f. diamminesilver(I) tetrachloroplatinate(II)

    g. sodium dicyanoargentate(I)   **h.** tetracyanonickelate(II) ion
    i. tetrabromocuprate(II) ion   **j.** tetraamminenickel(II) ion

Note: In part b, nickel is in the 0 oxidation state. Carbonyl is a neutral ligand. Therefore, the compound is neutral. This compound is dangerously poisonous. It was at one point considered to be the cause of "legionnaires' disease." The hypothesis was shown to be unfounded, however.

7.13. **a.** hexaaquamanganese(II) ion   **b.** tetracyanozincate(II) ion
    c. sodium hexachloroplatinate(IV)
    d. hexaamminecobalt(III) bromide   **e.** diamminesilver(I) nitrate
    f. hexafluorocobaltate(III) ion
    g. aluminum hexacyanoferrate(III)
    h. hexaaquamanganese(II) chloride
    i. potassium tetracyanonickelate(0)
    j. hexafluoroaluminate(III) ion

III.1.   **a.** proton donor   **b.** proton acceptor
      c. electron-pair acceptor   **d.** electron-pair donor

III.2.   **a.** HI   **b.** H_2O **c.** H^+ **d.** H_2O

III.3.   **a.** BF_3 **b.** F^-

III.4.   **a.** HF   **b.** HBr   **c.** HI   **d.** HCl   **e.** H_2CO_3 **f.** H_2SO_4
      g. H_2SeO_3 **h.** HIO_3

Note: The central atom of the oxyacid has the same oxidation number as the nonmetal in the acid anhydride.

III.5.   **a.** $HNO_3 + K_2SO_4$ **b.** $HBr + Na_2SO_4$
      c. $K_2SO_4 + HClO_4$ **d.** $HCl + K_2SO_4$
      e. $HClO_3 + CaSO_4$ **f.** $HBr + H_3PO_3$
      g. $HI + H_3PO_3$ **h.** $HCl + H_3PO_3$
      i. $HCl + H_3PO_4$ **j.** $HBr + H_3PO_4$

III.6.   **a.** KOH   **b.** $Sr(OH)_2$ **c.** RbOH   **d.** $Ba(OH)_2$
      e. $Al_2O_3 + H_2O$ **f.** $Fe_2O_3 + H_2O$ **g.** $SrO + H_2O$
      h. $K_2O + H_2O$

III.7.   **a.** $NaCl + H_2O$ **b.** $KBr + H_2O$ **c.** $CsI + H_2O$
      d. $Rb_2SO_4 + H_2O$ **e.** $BaCl_2 + H_2O$ **f.** $SrSO_4 + H_2O$
      g. $K_3PO_4 + H_2O$ **h.** $Ba_3(PO_4)_2 + H_2O$

III.8.

	Acid	Conjugate base
a. $H_3O^+ + CN^-$	HCN	CN^-
b. $H_3O^+ + ClO^-$	HClO	ClO^-
c. $H_3O^+ + F^-$	HF	F^-
d. $H_3O^+ + HSO_3^-$	H_2SO_3	HSO_3^-
e. $H_3O^+ + H_2PO_4^-$	H_3PO_4	$H_2PO_4^-$
f. $HCN + OH^-$	H_2O	OH^-
g. $HNO_2 + OH^-$	H_2O	OH^-

Note: In parts d and e, where polyprotic acids are undergoing ionization, the first ionization is given as the answer. You could, for example, show a second ionization for the hydrogen sulfite ion,

$$HSO_3^- \rightleftharpoons H^+ + SO_3^{2-}$$

III.9.   **a.** $HF + OH^-$ **b.** $HCN + OH^-$ **c.** $HNO_2 + OH^-$
       d. $NH_3 + H_3O^+$ **e.** $H_2CO_3 + OH^-$ **f.** $H_2PO_4^- + OH^-$
       g. $SO_4^{2-} + H_3O^+$ **h.** $HOAc + OH^-$

Note: In part g, acid is formed because HSO_4^- is the anion of the strong acid H_2SO_4. Thus there is no tendency for the HSO_4^- ion to accept a proton.

III.10.

	Lewis acid	Lewis base
a.	H^+	OH^-
b.	H^+	NH_3
c.	BF_3	F^-
d.	BF_3	NH_3
e.	SbF_5	F^-
f.	CO_2	O^{2-}
g.	Cu^{2+}	NH_3
h.	Zn^{2+}	OH^-

Note: In parts a and b, these reactions are also protonic acid-base reactions, but this fact is more obvious if we write H_3O^+ instead of H^+.

III.11.   **a.** $Cr(H_2O)_6^{3+}$ **b.** $Cu(NH_3)_4^{2+}$ **c.** $Ag(CN)_2^-$
       d. FeF_6^{3-} **e.** CrF_6^{3-} **f.** $Co(SCN)_4^{2-}$ **g.** $PtCl_4^{2-}$

III.12.   **a.** hydrobromic acid   **b.** hydrosulfuric acid
       c. orthophosphorus acid   **d.** hypochlorous acid
       e. chromic acid   **f.** (meso)arsenic acid
       g. (meta)selenous acid   **h.** (meta)aluminic acid
       i. hydriodic acid   **j.** chloric acid   **k.** hydrocyanic acid

III.13.   **a.** tetrachlorocuprate(II) ion   **b.** hexacyanoferrate(III) ion
       c. tetracyanozincate(II) ion   **d.** tetraiodocadmiate(II) ion
       e. dichloroargentate(I) ion   **f.** tetrachloroaurate(III) ion
       g. hexaamminecobalt(III) ion   **h.** hexaaquairon(II) ion
       i. diamminesilver(I) ion   **j.** tetraamminenickel(II) ion

CHAPTER 8

8.1.   **a.** Cu   **b.** Ag^+ **c.** Ag^+ **d.** Cu

8.2.   **a.** Na   **b.** Cl_2 **c.** Cl_2 **d.** Na   **e.** Cl   **f.** Na

8.3.

	(I)	(II)	(III)	(IV)
a.	Mg	Cl_2	Cl_2	Cl_2
b.	Cr	O_2	O_2	O_2
c.	Fe	H^+	H^+	H^+
d.	Na	Zn^{2+}	Zn^{2+}	Zn^{2+}
e.	Cu	NO_3^-	NO_3^-	NO_3^-

Note: Often the compound is called the oxidizing or reducing agent. For example, in part e, HNO_3 or NO_3^- may be referred to as the oxidizing agent when, in fact, it is the N atom that is reduced.

8.4.

	Higher	Lower	Type of agent
a.	6+	0	both
b.	none	0	oxidizing
c.	0, 4+, 6+	none	reducing
d.	4+, 6+	2−	both
e.	none	5+, 3+, 1+, 0, 1−	oxidizing
f.	none	0	oxidizing
g.	0	2−	both
h.	1−, 0	none	reducing

Note: Oxygen may have a + oxidation state, but only when it is combined with fluorine.

8.5.   **a.** N^{3-}, Cd, I^-, O^{2-} **b.** Al^{3+}, Mn^{7+}, F_2

8.6.   **a.** $NaCN + H_2$ **b.** $CrI_3 + H_2$ **c.** NR   **d.** NR
    e. $Zn(NO_2)_2 + H_2$ **f.** $KF + H_2$ **g.** NR   **h.** $SnCl_2 + H_2$
    i. $CdSO_3 + H_2$ **j.** $Mg_3(PO_4)_2 + H_2$

8.7.   **a.** $Ca(OH)_2 + H_2$ **b.** $Zn(OH)_2 + H_2$ **c.** NR   **d.** NR

e. $Cr(OH)_3 + H_2$ **f.** NR   **g.** $Sr(OH)_2 + H_2$ **h.** NR
i. $KOH + H_2$ **j.** NR

8.8. **a.** reducing, loses   **b.** oxidizing, reduced   **c.** reducing
d. loses, increases

8.9. **a.** $LiCl + Zn$ **b.** NR   **c.** NR   **d.** $MnCl_2 + H_2$
e. NR   **f.** NR   **g.** $KCN + Ag$ **h.** $PbBr_2 + H_2$
i. NR   **j.** $SrBr_2 + Ni$

8.10. **a.** $BaCl_2 + Sr$ **b.** NR   **c.** $AgCN + Au$ **d.** NR

8.11. **a.** S and N are in high oxidation states and therefore can be reduced.
b. They are already in their lowest possible oxidation states. To be oxidizing agents, they would have to be reduced, and this is not possible.

8.12. **a.** $AgNO_3 + NO$ (or NO_2) $+ H_2O$ **b.** $Cr_2(SO_4)_3 + H_2$
c. $HgSO_4 + SO_2 + H_2O$ **d.** NR

8.13. **a.** $Al(NO_3)_3 + NH_3 + H_2O$ **b.** $Sn(NO_3)_2 + NO + H_2O$
c. $SnSO_4 + SO_2 + H_2O$ **d.** $HgSO_4 + SO_2 + H_2O$
e. $K_2SO_4 + S_8 + H_2O + H_2S$ **f.** $Rb_2SO_4 + H_2$

Note: In part e both S_8 and H_2S would likely be formed. It is also possible that in part a, some N_2 will be formed, and in part b N_2O or N_2 might also be formed. The actual products will be dependent on the conditions of the reaction.

8.14. **a.** $KCl + CrCl_3 + Cl_2 + H_2O$
b. $Cu(NO_3)_2 + S_8 + NO + H_2O$ (or SO_2)   **c.** $SO_2 + H_2O$
d. $NO + S_8 + H_2O$ **e.** $MnSO_4 + Br_2 + K_2SO_4 + H_2O$

Note: It is reasonable to expect other reduction products such as NO_2 and N_2 in parts b and d, and perhaps some SO_2 rather than or along with S_8 in part d.

8.15. **a.** $U^{3+} + Cr^{3+} \longrightarrow U^{4+} + Cr^{2+}$
b. $Cu^+ + Fe^{3+} \longrightarrow Cu^{2+} + Fe^{2+}$
c. $Sn^{2+} + Co^{3+} \longrightarrow Sn^{4+} + Co^{2+}$

8.16. **a.** $F_2 + 2\,Cl^- \longrightarrow 2\,F^- + Cl_2$
b. $Br_2 + 2\,I^- \longrightarrow 2\,Br^- + I_2$
c. $H_2 + Cl_2 \longrightarrow 2\,H^+ + 2\,Cl^-$

8.17. **a.** $BaCl_2 + Zn$ **b.** NR   **c.** $NiBr_2 + H_2$ **d.** NR
e. $NaOH + H_2$ **f.** NR   **g.** NR   **h.** NR   **i.** $Hg(CN)_2 + Ag$
j. NR   **k.** $Mg(NO_3)_2 + NH_3 + H_2O$ **l.** $PbSO_4 + SO_2 + H_2O$
m. $Sn(ClO_4)_4 + Cl_2 + H_2O$ **n.** S_8 (or SO_2) $+ NO + H_2O$
o. $MnCl_2 + SnCl_4 + H_2O + KCl$ **p.** $Ag + Fe^{3+}$ **q.** NR
r. $Cr^{3+} + H_2$ **s.** $I_2 + Cl^-$

CHAPTER 9

9.1. **a.** direct-combination or redox   **b.** redox
    c. metathesis   **d.** decomposition
    e. direct-combination (acid anhydride and water)
    f. direct-combination (basic anhydride and water)
    g. protonic acid–hydroxyl base   **h.** Brønsted-Lowry acid-base
    i. direct-combination (Brønsted-Lowry or Lewis acid-base)
    j. direct-combination (Lewis acid-base)

9.2. **a.** Cu^{2+} **b.** Ag^{+} **c.** Al^{3+} **d.** Zn^{2+}
    e. To be an oxidizing agent, Cl^{-} must be reduced. However, it is already in its lowest oxidation state and cannot be reduced.

9.3. $AuCl_4^{-}$, tetrachloroaurate(III) ion

9.4. **b.** H_2CO_3, $4+$, $4+$ **c.** H_2SO_4, $6+$, $6+$ **d.** H_3PO_3, $3+$, $3+$
    e. H_3PO_4, $5+$, $5+$

9.5. **a.** $Na_2SO_4 + HCl$ **b.** $AgNO_3 + NO$ (or NO_2) $+ H_2O$
    c. $AuCl_4^{-} + NO + H_2O$ **d.** $SO_3 + NO$ **e.** H_2SO_4
    f. $NO + H_2O$ **g.** NO_2 **h.** $HNO_3 + NO$

9.6. **a.** $Cu_2O + SO_2$ **b.** $HNO_3 + NO$ **c.** $Sn + CO$
    d. $Na_2SO_4 + HNO_3$ **e.** $CaO + CO_2$ **f.** $PbO + SO_2$
    g. $Cu + SO_2$ **h.** $Zn + CO$ **i.** H_2SO_3 **j.** $Pb + CO_2$

9.7. **a.** reducing   **b.** Zn   **c.** Zn

9.8. **a.** dicyanoargentate(I) ion   **b.** tetrachloroaurate(III) ion
    c. tetracyanozincate(II) ion

9.9. **a.** $Cu_2O + SO_2$ **b.** $Ag + Zn(CN)_4^{2-}$ **c.** $HI + Na_2SO_4$
    d. $AuCl_4^{-} + NO + H_2O$ **e.** $Cu(NO_3)_2 + NO + H_2O$

9.10. **a.** $3+$ **b.** $2\frac{2}{3}+$ **c.** $2+$

9.11. **a.** H_2SiO_3 **b.** $HAlO_2$

9.12. **a.** $Fe_3O_4 + CO_2$ (or $Fe + CO_2$)   **b.** $CaSiO_3$ **c.** $HBr + Na_2SO_4$
    d. $Zn(NO_3)_2 + NH_3 + H_2O$ (NH_4NO_3)   **e.** $Ag + Zn(CN)_4^{2-}$
    f. $Fe + CO_2$ **g.** HNO_2 **h.** $HNO_3 + NO$ **i.** NH_3
    j. $Sn + CO$

Note: In part d, if sufficient HNO_3 is present, the NH_3 will react with the HNO_3 to form NH_4NO_3.

9.13. a. $CaO + CO_2$　　**b.** $Ca(OH)_2$　　**c.** $CaCO_3 + H_2O$

9.14. basic anhydride, H_2CO_3

9.15. acid, base

9.16. a. $CaO + CO_2$　　**b.** $H_2O + O_2$　　**c.** $H_2O + CO_2$
　　　d. $Na_2CO_3 + H_2O + CO_2$　　**e.** $CO_2 + H_2O$　　**f.** $CO_2 + H_2O$

9.17. $H_2O + CO_2$

9.18. $CO_2 + H_2O$

9.19. a. $Ca(OH)_2$　　**b.** $CaCO_3 + H_2O$　　**c.** $Na_2CO_3 + H_2O + CO_2$
　　　d. $H_2O + CO_2$　　**e.** $AuCl_4^- + NO + H_2O$　　**f.** H_2SO_4
　　　g. $HNO_3 + NO$　　**h.** $Sn + CO$　　**i.** $Cu_2O + SO_2$　　**j.** $AgCl_2^-$
　　　k. $Fe_3O_4 + CO_2$ (or $Fe + CO_2$)　　**l.** $CaSiO_3$　　**m.** $H_2O + CO_3^{2-}$

9.20. BrO_3^-, Br^-

9.21. BrO_3^-, Br^-

9.22. a. $I_2 + Cl^-$　　**b.** $I_2 + Mn^{2+} + H_2O$　　**c.** $Br_2 + Mn^{2+} + H_2O$
　　　d. $I_2 + Cr^{3+} + H_2O$

9.23. a. $F^- + Br_2$　　**b.** $Cl_2 + Mn^{2+} + H_2O$　　**c.** NR
　　　d. $I_2 + Mn^{2+} + H_2O$　　**e.** $Br_2 + H_2O$

9.24. a. $HF + CaSO_4$　　**b.** $HNO_3 + NO$　　**c.** H_2SO_4
　　　d. $Ag(CN)_2^- + Cl^-$　　**e.** $CuO + CO_2$　　**f.** $HNO_3 + Na_2SO_4$
　　　g. $Cu + CO_2$　　**h.** $CaCO_3 + H_2O + CO_2$
　　　i. $I_2 + H_2O + SO_4^{2-} + H^+$　　**j.** $Br_2 + Cl^-$
　　　k. $Pb + CO_2$　　**l.** $Mn^{2+} + Cl_2 + H_2O$

9.25. a. $S_8 + H_2O$　　**b.** $NaOAc + H_2O + CO_2$　　**c.** $Ca(OH)_2$
　　　d. $CaSiO_3$　　**e.** $ZnO + SO_2$　　**f.** $CaO + CO_2$　　**g.** H_2CO_3
　　　h. $H_2O + SO_2$　　**i.** $I_2 + Mn^{2+} + H_2O$　　**j.** Na_2SO_4
　　　k. $Sn + CO$　　**l.** $Zn(CN)_4^{2-} + Ag$　　**m.** $Cr^{3+} + Cl_2 + H_2O$